Advances in Industrial Control

Other titles published in this series:

Jacques Richalet • Donal O'Donovan

Predictive Functional Control

Principles and Industrial Applications

With a Foreword by Karl Åström

 Springer

Jacques Richalet, PhD
3 allée des Nourets
78430 Louveciennes
France
jacques.richalet@wanadoo.fr

Donal O'Donovan
Cork Institute of Technology
Department of Electronic Engineering
Rossa Avenue
Bishopstown, Cork
Ireland
donal.odonovan@cit.ie

ISSN 1430-9491
ISBN 978-1-84996-845-4   e-ISBN 978-1-84882-493-5
DOI 10.1007/978-1-84882-493-5
Springer Dordrecht Heidelberg London New York

British Library Cataloguing in Publication Data
A catalogue record for this book is available from the British Library

Cover design: eStudioCalamar, Figueres/Berlin

Printed on acid-free paper

Springer is part of Springer Science+Business Media (www.springer.com)

Advances in Industrial Control

Series Editors

Professor Michael J. Grimble, Professor of Industrial Systems and Director
Professor Michael A. Johnson, Professor (Emeritus) of Control Systems and Deputy Director

Industrial Control Centre
Department of Electronic and Electrical Engineering
University of Strathclyde
Graham Hills Building
50 George Street
Glasgow G1 1QE
United Kingdom

Series Advisory Board

Professor E.F. Camacho
Escuela Superior de Ingenieros
Universidad de Sevilla
Camino de los Descubrimientos s/n
41092 Sevilla
Spain

Professor S. Engell
Lehrstuhl für Anlagensteuerungstechnik
Fachbereich Chemietechnik
Universität Dortmund
44221 Dortmund
Germany

Professor G. Goodwin
Department of Electrical and Computer Engineering
The University of Newcastle
Callaghan
NSW 2308
Australia

Professor T.J. Harris
Department of Chemical Engineering
Queen's University
Kingston, Ontario
K7L 3N6
Canada

Professor T.H. Lee
Department of Electrical and Computer Engineering
National University of Singapore
4 Engineering Drive 3
Singapore 117576

Professor (Emeritus) O.P. Malik
Department of Electrical and Computer Engineering
University of Calgary
2500, University Drive, NW
Calgary, Alberta
T2N 1N4
Canada

Professor K.-F. Man
Electronic Engineering Department
City University of Hong Kong
Tat Chee Avenue
Kowloon
Hong Kong

Professor G. Olsson
Department of Industrial Electrical Engineering and Automation
Lund Institute of Technology
Box 118
S-221 00 Lund
Sweden

Professor A. Ray
Department of Mechanical Engineering
Pennsylvania State University
0329 Reber Building
University Park
PA 16802
USA

Professor D.E. Seborg
Chemical Engineering
3335 Engineering II
University of California Santa Barbara
Santa Barbara
CA 93106
USA

Doctor K.K. Tan
Department of Electrical and Computer Engineering
National University of Singapore
4 Engineering Drive 3
Singapore 117576

Professor I. Yamamoto
Department of Mechanical Systems and Environmental Engineering
The University of Kitakyushu
Faculty of Environmental Engineering
1-1, Hibikino,Wakamatsu-ku, Kitakyushu, Fukuoka, 808-0135
Japan

Series Editors' Foreword

The series *Advances in Industrial Control* aims to report and encourage technology transfer in control engineering. The rapid development of control technology has an impact on all areas of the control discipline. New theory, new controllers, actuators, sensors, new industrial processes, computer methods, new applications, new philosophies..., new challenges. Much of this development work resides in industrial reports, feasibility study papers and the reports of advanced collaborative projects. The series offers an opportunity for researchers to present an extended exposition of such new work in all aspects of industrial control for wider and rapid dissemination.

The 1960s witnessed many changes in social, economic, and technological activities. In the theory of control systems, a paradigm change was taking place, and the state-space approach, optimal control, and the Kalman filter are just three concepts that emerged around that time that have had a lasting impact on the subsequent development of system analysis and control systems design. Industrial control systems technology was also changing, responding to industrial demands for better performance from systems of increasing complexity.

The potential of direct digital control systems and the computer control of processes was being explored to provide better co-ordinated control and improvements in process control and product quality. However, the end of the decade witnessed the emergence of the microprocessor and by 1971 the first Intel® microprocessor chip was on the market. The new capabilities offered by these computer-based control systems meant that control engineers could begin to control larger systems using more advanced control algorithms that were embedded in well-designed hierarchical process control structures. The petrochemicals, steel, and utility industries were among the first industries to use and exploit these new capabilities.

Motivated by the needs of the petrochemicals and similar industries, Jacques Richalet began to develop and promote new predictive control ideas during the late 1960s and early 1970s. In fact, Jacques Richalet is often regarded as the "grandfather" of model predictive control (MPC). He formulated his ideas as the predictive functional control (PFC) algorithm and over the next decade began the first industrial applications of his new method (distillation/reactors/furnaces). His first industrial application of MPC was in 1973. A key motivation was to provide better performance than could be obtained with the widely-used PID controller whilst making it easy to replace the PID controller unit or module with his new algorithm. It was the advent of digital control technology and the use of software control algorithms that made this replacement easier and more acceptable to process engineers.

A decade of industrial practice with PFC was reported in the archival literature by Jacques Richalet *et al.* in 1978 in an important seminal *Automatica* paper. Around this time, Cutler and Ramaker published the dynamic matrix control algorithm that also used knowledge of future reference signals to determine a sequence of control signal adjustment. Thus, the theoretical and practical development of predictive control methods was underway and subsequent developments included those of generalized predictive control, and the whole armoury of MPC methods.

Jacques Richalet's approach to PFC was to seek an algorithm that was:

- easy to understand;
- easy to install;
- easy to tune and optimise.

He sought a new modular control algorithm that could be readily used by the control-technician engineer or the control-instrument engineer. It goes without saying that this objective also forms a good market strategy.

Jacques Richalet and Donal O'Donovan have adopted a similar approach in this *Advances in Industrial Control* monograph entitled *Predictive Functional Control*, with a strong emphasis on industrial nomenclature and applications. It is a multi-layered book that will appeal to a wide variety of readers. The book can be read by those who seek a general overview of the potential of PFC or by those who have a good theoretical and practical knowledge of control and wish to study the technique in depth. Richalet and O'Donovan have included new material on PFC that has not been published before. In the opening pages of the book, the authors have thoughtfully provided a useful Reading Guide to the content, purpose and technical level of the individual chapters. This should help readers find their way through the book easily.

We are fortunate to be able to welcome this book on PFC into the *Advances in Industrial Control* series for it is not often we are able to include a book that benefits from over forty years of development knowledge and applications experience from one of Europe's leading control engineers.

Industrial Control Centre   *M.J. Grimble*
Glasgow   *M.A. Johnson*
Scotland, UK
2008

Foreword

The first time I heard about model predictive control (MPC) was in a lecture given by Jacques Richalet at the IFAC Symposium on System Identification in Tiblisi in 1976. Richalet not only presented the key ideas, but he also presented several industrial applications. Since the ideas were not conventional it took some time before they were first published in an *Automatica* paper in 1978. The concept of MPC has been rediscovered several times; most notably by Cutler and Ramacher at Shell, who were also motivated by practical problems. Interest in MPC has accelerated significantly over the last 20 years, driven by the dramatic increase in computing power and a strong industrial interest. MPC is still, more than 30 years after its conception, of very high interest, both in academia and industry. It occupies 2nd, 3rd and 6th places among the most cited papers in *Automatica*.

Richalet has continued his interest in MPC and has applied his technique in numerous applications. He has also disseminated his knowledge though training courses. It is a real pleasure to have a book written by the originator of MPC, and one of his collegues, Donal O'Donovan. This book, which is intended for a wide audience, requires only a modest theoretical background; block-diagram algebra, Laplace- and z-transforms, and PID control. The book presents the basic concepts, and contains much information about the practical details, such as implementation and tuning.

This book should prove very useful to practical engineers who are interested in moving beyond PID control, and to theoreticians who would like to get a flavour of real industrial problems. I am particularly pleased that it emphasises the importance of transparent specifications and the usefulness of having performance-related knobs on a controller.

Lund, January 2009   *Karl Johan Åström*

Preface

Intended Audience

This book is intended for people with many different backgrounds and objectives.

Category 1

Experienced individuals with a technical background: Engineers and senior technicians from varied engineering disciplines such as mechanical, chemical, electrical, *etc.*, possessing a limited experience of automated systems. Such people would be seeking background information rather than an in-depth analysis in model-based predictive control techniques.

Category 2

Individuals educated in automation: People who have been exposed to the concepts of a process, transfer function, stability, *etc.*, but have not had the opportunity to put this theory into practice. Also, it is intended for those who wish to be kept informed of developments in industrial automation without going into the practical implementation details.

Category 3

Automation industry practitioners: PID specialists, instrumentation personnel and those familiar with the issues surrounding instrumentation, actuators and sensors, which are required to implement continuous controllers. It is also intended for people with a long-term requirement for a more effective form of control and who are seeking a guide for potential controllers.

Category 4

Individuals who have attended seminars or graduate courses in "advanced control": Those who possess a good theoretical and practical background in automatic control and make use of controllers other than PID.

The use of mathematical tools is kept to a minimum in order to retain the emphasis on the concepts. The level of mathematics used is equivalent to that of a junior engineer.

However, a basic understanding of systems theory would be beneficial. This would include a basic understanding of fundamental concepts such as:

- block-diagram algebra;
- the Laplace and z-transforms for the continuous and discrete time domains (although, this is not crucial);
- elementary knowledge of PID.

Reading Guide

Chapters 1, 2 and 3 present the basic principles and fundamental concepts of the predictive control methodology in an intuitive manner.

Chapter 4 presents a mathematical derivation of the predictive regulator in its most basic form. Extensions to the basic regulator are then proposed for controlling integrating, time-delayed and unstable systems. A basic knowledge of the mathematical tools mentioned in Category 4 above would be useful.

Chapter 5 discusses the factors that influence the tuning of the predictive regulator. The chapter concludes by presenting a tuning aid that suggests how the regulator parameters should be altered when particular results are obtained.

Chapter 6 describes how the presence of the system constraints may be dealt with by imposing limits on the range of various parameters of the predictive regulator and through the use of logical supervisors.

Chapter 7 puts the concepts described in the previous chapters into practice. Many control structures are discussed, some of which may be used in conjunction with a PID regulator. This chapter demonstrates techniques of practical benefit in industrial applications and *should be read by everyone*, if for no other reason than to become familiar with the specific control structures involved.

Chapter 8 is aimed specifically at thermal process control specialists. The previously unpublished material contained within this chapter is of great practical benefit.

Chapter 9 deals with unstable and non-minimum phase (inverse unstable) systems, which are difficult to control and fortunately rare. Consequently, this chapter does not constitute essential reading for all.

Chapter 10 describes several industrial applications. These examples are provided to demonstrate that predictive control is far more than just an academic concept. It is felt that the examples chosen are of interest generally, rather than confining the discussion to specialist areas – although such discussions do contain merit.

Chapter 11 discusses important precautions to be taken when using the predictive control technique and describes the associated risks.

Some chapters, such as Chapters 8 and 9, contain original material that, although having been applied in practice, has not been presented previously. By its nature, the novel content of these chapters is more demanding than that of the initial introductory and final assessment chapters.

The following guide indicates the suggested reading path for each category of reader.

Chapter		Category			
		1	2	3	4
1	Why Predictive Control?	*	*	*	*
2	Internal Model	*	*	*	*
3	Reference Trajectory	*	*	*	*
4	Control Computation	-	-	*	*
5	Tuning	-	-	*	*
6	Constraints	-	-	*	*
7	Industrial Implementation	-	*	*	*
8	Parametric Control	~	~	*	*
9	Unstable Poles and Zeros	-	-	-	~
10	Industrial Examples	*	*	*	*
11	Conclusions	*	*	*	*

*	To read
-	May be omitted during first reading
~	Personal choice

Acknowledgments

The authors would like to thank Prof. Karl Åström, who was the first to acknowledge the potential of the model-based predictive control technique and supported our efforts from that time on. We would also like to thank Prof. Michael Grimble, Prof. Michael Johnson and Larry Poland for their advice and support during the editing of this book. Also, we would like to express our gratitude to Joëlle Mallet and Guy Lavielle for their fundamental contribution to the industrial applications.

Many thanks go to the following for their contributions during the writing of this book: Prof. Robert Haber, Dr. Tom O'Mahony, Prof. John Ringwood and Dr. Derry Fitzgerald.

We would like to express our sincere gratitude to the many end-users for their insightful remarks, criticisms and suggestions that have contributed greatly to the development of the techniques presented in this book: YPF (Ar), Repsol (Sp), Lubrisol (F), Elf ATo (F), Sofiproteol (F), SollacArcellorMittal (F), Total (F), CNR (F), Evonik.Degussa (D), Thomson (F), Pechiney Alcan (F), Renault (F),

Peugeot PSA (F), ESA (Eu), SanofiAventis (F), EDF (F), SNCF (F), Mitsui-chem (JP), JGC (JP), CSEE (F), Lactalis (F), Tour Eiffel (F), Schneider Electric (F), Nodon (F), Veolia (F), Sepro (F), SFIM (F), Marine Nationale, Armée de l'air (F).

Contents

Abbreviations and Symbols

The significance of some symbols may vary from chapter to chapter. In such circumstances, the interpretation adopted will be stated explicitly.

A	Heat-exchange surface (m^2)
a_m	Control decrement of first-order model
a_p	Control decrement of first-order process
CLTR	Closed-loop time response
Cons	Set-point (interchangeable with Setpoint)
Cp	Specific heat capacity ($J\,K^{-1}\,kg^{-1}$)
DM	Delay Margin
CV	Controlled variable (interchangeable with PV)
DV_m, DV	Measured disturbance
DV_{nm}	Unmeasured disturbance
ε	Error
e_m	Model input
e_p	Process input
h_s	Coincidence point
$H(s)$	Continuous transfer function
GM	Gain Margin
K_m	Model steady-state gain
K_p	Process steady-state gain
λ	Reference trajectory decrement
l_h	Error multiplier
M_N	N^{th} model transfer function
MV	Manipulated variable
n	Sample index
NUT	Number of transfer units
OLTR	Open-loop time response
P	Process transfer function
PME	Measured process and model predicted output error
PV	Process variable (interchangeable with PV)
r	Pure discrete time delay

R	Regulator
P	Density ($kg\ m^{-3}$)
s	Laplace transform variable
Setpoint	Set-point (interchangeable with Cons)
SISO	Single-input, single-output system
s_m	Model output
s_p	Process output
t	Time
T, τ	Time constant
T_s, T_{ech}	Sampling period
U	Heat-exchange coefficient
ω	Frequency ($rad\ s^{-1}$)
z	z-transform variable

1

Why Predictive Control?

Abstract. This chapter begins by discussing the routine activity of driving a car. The discussion serves to highlight how people regularly and naturally apply the principles of model-based predictive control on a daily basis. The historical context that gave rise to the Predictive Functional Control (PFC) technique, the subject of this book, is briefly presented and illustrates why and how an alternative to the proportional, integral, derivative (PID) controller was, and still is, required for demanding processes. The chapter concludes by presenting a fundamental block-diagram framework capable of representing the processes utilised throughout this book.

Keywords: PFC, PID, block diagram

1.1 "You would not drive your car using PID control"

To introduce the concepts of predictive control we begin by considering the common, everyday activity of driving a car. When driving, the ultimate goal is to reach the destination on time and in a safe manner. A driver achieves this goal by subconsciously adopting a series of intermediate objectives over a fixed distance or horizon that recedes[1] with the passage of time ahead of the vehicle. These objectives take the form of speed profiles or trajectories, that are chosen to satisfy the desired driving requirements, *e.g.*, is the driver in a hurry or out for a leisurely drive? In practice, variable road and weather conditions, such as rain, wind, slope, oil patches, *etc.*, introduce unknown factors or disturbances that must be constantly monitored by the driver and compensated for by making the appropriate corrections to the speed and direction of the car.

---

[1] The term "recedes" is used to convey the effect observed as the speed trajectory is re-assessed following the application of each new speed correction or control effort, *e.g.*, accelerator adjustment. This modified profile effectively constitutes a window or horizon that shifts or "slides" in time.

Personal experience and training has taught us that any corrections made using the accelerator, brakes or steering will subsequently affect the speed and position of the car to varying degrees. A complete understanding of the underlying mechanical principles involved is not required for these operations to be successful. What is required is knowledge of the relationships, or a mental *model*, between the methods of speed and steering correction and their associated effects on the vehicle. This information allows the driver to select the most appropriate adjustment mechanism(s) (*i.e.*, set-point adjustment) for the vehicle's speed.

Consider the more realistic situation where the vehicle's inertia is taken into account. Under such circumstances, delaying the application of the brakes until the vehicles are too close would result in catastrophic consequences.

Experience has taught us that it is necessary to anticipate the behaviour of other vehicles (and obstacles) *far* in advance to prevent accidents. Such knowledge reinforces the individual connections made in a driver's brain between a particular action and its related effect.

The converse may also be learned and put into practice, *i.e.,* that the desired behaviour is associated with a particular action – a fact so fundamental to our everyday existence that it is sometimes taken for granted and its potential benefits may be overlooked.

To ensure success, the effectiveness of a particular action should be observed in order to verify that the outcome is as expected. If this proves not to be the case, *i.e.*, that the observed behaviour diverges from the predicted behaviour, this knowledge should then be used to alter the driver's mental model to reflect the vehicle's true behaviour more accurately.

But some care needs to be taken when applying such an approach. The possibility also exists that the discrepancy between the vehicles' *observed* and *predicted* behaviours may be attributable to some random, temporary disturbance – a point that will be addressed at a later stage.

The simple example above highlights the fundamental principles and requirements of predictive control, which are:

- it should be possible to represent the process response to a given stimulus using some form of mathematical model;
- a projection of the desired process behaviour should be expressed in the form of an "ideal" or reference trajectory;
- the action is dictated by the relation: **Action → Effect**, from which the Desired Effect → Applied Action is determined;
- possible modification of the control system's internal model.

1.2 Historical Context

In the mid-1950s, the operational amplifier, with all its improvements – higher gain/bandwidth, less drift, temperature compensation, high reliability, *etc.* became available at an affordable price. This was a significant improvement on the previous technology.

Concurrently, the theoretical methodology and stability problems associated with feedback control systems were well understood, solved and implemented in practice with servomechanisms.

The replacement of the old technologies, *i.e.*, valves, pneumatics, *etc.*, by the new electronic controller and advances in control techniques was at the heart of the remarkable proliferation of PID in industry at that time.

At the end of the 1960s, the digital computer became reliable, easily programmable and cheap. So, a greater responsibility could be given to this new control device. On the other hand, static and dynamic modelling, simulation and identification techniques were maturing – partly due to the fact that scientific computation was now feasible.

The digital computer brought together, for the first time, memory and algorithmic execution, which are a mixture of logic and the simulation of dynamic systems. The merging of technology and methodology thus gave birth to the concept of model-based control strategy being implemented in a digital computer.

The impetus arising as a consequence of this merging was such that it was inevitable that the same basic concept (computer, model and prediction) would occur to and develop in different minds at different locations. Every time there is a concurrent leap in technology and methodology, there is a significant improvement in industrial practice.

1.3 Breaking with the PID Tradition

PID control is one of the oldest and most common control techniques used in industry today. It long preceded the introduction of digital computer control, which has revolutionised the automation industry. Although, in many respects, predictive control represents a departure from the PID control philosophy, it does not represent a conflict in methodology. In fact, the application areas of both approaches are complementary. Each strategy possesses its own strengths and weaknesses. The practitioner, using knowledge and experience, must choose the control scheme that is more appropriate for the problem at hand.

Since the introduction of the PID regulator, great advances have been made in the fields of computer technology and mathematical modelling techniques. The introduction of such computer-based simulation/design tools has re-defined the performance expectations of modern systems.

Ultimately, both control strategies are implemented in software as a set of equations that, in some cases, may be quite similar. However, there are fundamental differences between the two techniques at the first principles, commissioning and tuning stages. The manner in which process behavioural information is utilised best illustrates this point. Predictive control uses a model and measurements, while PID control bases its decisions on measurements only. For predictive control to be effective it is necessary to possess *knowledge* of the process.

This information is contained within the two components, measurements and knowledge. Both are necessary, but each is used to differing degrees:

- **Measurements:** derived from sensors for position, speed, temperature, flow, *etc.*
- **Knowledge of the process structure:** such knowledge permits prediction. For example, a driver understands that applying the brakes produces a deceleration. From a predictive control perspective, such information may be incorporated in the controller by using a mathematical model.

At the time of the PID controller's inception, computers capable of implementing modelling algorithms did not exist. Thus, the only source of information available was from sensor measurements; this is referred to as "state information". The only feasible strategy available was to generate a control signal based on a "universal" law that was independent (or almost) of the underlying process to be controlled. This law derived its output as a function of the observed sensor measurements.

The general technique involves summing various forms of the error signal, *i.e.,* the difference between the measured magnitude and its set-point value, using appropriate weighting coefficients of the error signal, its integral and derivative to generate the controller output. These weightings are chosen as a function of the process to be controlled and, from a certain perspective, may be viewed as a means of "modelling" the plant within the controller.

However, the output of a process also depends on unknown input disturbances (*i.e.,* road and weather conditions) and the past actions of the controller (driver behaviour). Since most processes are low-pass in nature the output of the process depends, in a predictable way, on what the controller has generated in the past. So, it would be natural to take the actual position of the car on the road, measured by some sensor, and the predictable behaviour arising from the past actions of the driver into account

Thus, the decision to alter the control settings of the car (*e.g.,* steering wheel position, accelerator) is based on observations of the current driving situation (current sensory feedback) and a prediction of future car behaviour. The prediction of future behaviour is founded on the driver's past familiarity with the car's response to the known actions of the driver (*i.e.,* a mental model of the car's performance).

It may be argued that the integral term partly contains the past actions of the driver. Yes, this is true to a certain extent. But, it is not possible to discriminate between the known, past actions of the driver and the natural behaviour of the car using the integral term alone. This is so because it is not possible to distinguish between the predictable actions of the driver and the perturbations acting on the car (*e.g.,* slope, wind, road conditions).

As with any process, the car's response to change in input consists of two components: an open-loop component, consisting of the effects of past actions and a closed-loop component, which is used to counteract disturbances. The open-loop component cannot be influenced and will occur, irrespective of the driver's future actions. However, the closed-loop component is dictated by the current and future decisions made by the driver.

Model-based predictive control takes what is predictable in the future into account and controls the closed-loop component to achieve the desired response. It is clear that isolating what may be controlled from the natural, open-loop behaviour of the process simplifies the whole control procedure and, as a consequence, improves the results.

The effects of measured and unmeasured variables on the process are not independent. They are interconnected by what is referred to as the *structure*, which is assumed to be invariant (or at least remains constant while the control signal is being applied) and may be represented by a mathematical model.

Now, assume that we have complete understanding of the process behaviour and that we can control all the relevant environmental factors. Such a situation gives rise to *perfect structural information*. Under these circumstances, measuring the regulated output is not necessary as control of the process may be achieved by merely calculating the inverse of the *Action* → *Effect* relationship. The implicit assumption made here is that a perfect model of the process is known and may be trusted completely. Of course, such an assumption is unrealistic in reality!

A common question that arises is "One can perform miracles if the process model is perfect, but what happens if it's not?" This is an important issue and it is often used as an argument to prevent advocates of predictive control from becoming over-confident in its abilities!

Clearly, the solution lies between the following two extreme strategies:

- **Perfect structural information available:** In such circumstances open-loop control may be applied using the inverse of the process model without any fear of stability issues.
- **Absence of any structural information:** Closed-loop control may be employed using an optimal combination of signal measurements.

As models are always *inaccurate*, predictive control endeavours to find a local compromise between these two extreme strategies. This closed-loop approach achieves its goal by utilising a prediction of the system output based on an internal model of the process.

If the process is too *complex* to be modelled (we will not define what complex means) an appropriate combination of signals may be passed to a rule-based control system to perform the regulation. If the process is well understood, the number of control commands required may be reduced to a minimum. This is the approach adopted by humans when driving a vehicle. The appropriate channels of information, such as the observation of relative motion, depth perception, *etc.*, may be combined with reasoning, knowledge and experience to compensate for any missing information. It is this *minimalist* approach to controlling their environment that gives humans the ability to perform numerous tasks simultaneously.

Consequently, the fundamental principles of predictive control are easy to understand because they originate from human behaviour. In fact, it is a form of control that may be considered as "natural".

1.4 Impact on Industry

The principles of predictive control date from the end of the 1960s; whereas the first recorded industrial applications date from the early 1970s. For various reasons, the petrochemical and defence industries were first to show interest and offer financial support for the development of the technique. Predictive control has several interesting characteristics but, historically, there were two specific characteristics that played a defining role. These were constraint handling and zero tracking (or lag) error.

1.4.1 The Petrochemical Industry

The oil crisis of the 1970s saw the operational profit margins of oil refineries dwindle and consequently optimisation became a necessity as opposed to a luxury. In effect, optimisation consists of operating the various installed units at different technical limits (physical constraint, *e.g.*, distillation, column flooding) while utilising the maximum range of the actuators (technical constraint, *e.g.*, maximum valve opening).

Dynamic regulation demands that process control and internal state variable *constraints* must be taken into account at all times. In Chapter 6, we will see that predictive control can naturally account for these two types of constraints in a simple and satisfactory manner.

Constrained multivariable control was introduced to tackle the problems associated with large-scale distillation where the required return time on investment is very short. Such systems necessitated large financial budgets and were available only from specialised automation companies. These ventures proved quite profitable for practitioners and vendors alike, and propagated quickly throughout the oil industry. Today, there are thousands of such installations.

1.4.2 The Defence Industry

The most common problems faced by weapons control systems involve tracking a moving target. The ability to follow a set-point with zero tracking (or lag) error will be considered in Chapter 5 and was the primary characteristic that attracted the attention of the defence industries.

Numerous non-defence-related applications ensued as a result. Initially, the robotics industries adopted the approach and were soon followed by the chemical industry for use in heat treatment applications where it is necessary to follow a temperature profile. Other industrial sectors followed.

In the pharmaceutical industry, it is crucial for the crystallisation process to follow time varying set-points accurately.

1.5 Objective

The application areas described in the previous sections may be notionally described as "expensive", as few financial budgetary constraints applied. However, is it possible to make predictive control accessible to floor instrumentation personnel in the traditional industries? Can predictive control be presented in a form such that it may be used to quickly control systems that have proved difficult for PID in the past, thus adding another tool to the practitioner's toolbox? The answer to this question is "definitely yes!"

The factors that determine the practical success of a controller are its economic cost and operational philosophy, *i.e.*, computer implementation and commissioning. Any controller that is to be presented as a credible alternative to PID must be comparatively cheap to implement and commission and relatively easy to understand if it is to stand any real chance of being accepted by industry.

Difficulties with constraints and zero-order tracking are not only encountered in non-linear, multivariable processes. For example, consider the case of a simple, first-order, level-regulated control system with a large time delay and possessing supplementary dynamics due to the sensors and actuators. In this situation, a PID controller would prove unsatisfactory for all but the most basic regulation requirements. Tightening the requirements would, long before the onset of instability, produce an oscillatory control signal that would result in a reduction in the lifetime of the actuator − a valve being a typical example.

It is not being suggested that predictive control should be substituted for all other forms of control. As with any algorithm, it too has its strengths and weaknesses. Its most valued characteristic is its *ease of tuning*. Thanks to this feature, predictive control is now widely implemented in industrial control systems.

Thus, the objective is to design a predictive controller that incorporates the benefits of the characteristics outlined above while minimizing any adverse effects that these factors might have on algorithm performance. Three conditions must be fulfilled if this objective is to be satisfied [1,5,7].

1.5.1 Formulation

Any method, whatever its merits, is only as good as the expertise that implements it. Thus, to ensure a good foundation of understanding, the mathematical formulation to be adopted should be accessible to those who are new to the profession, not only to the experienced practitioners. The mathematical framework used must also be capable of representing the process at a sufficient degree of complexity while remaining understandable to as broad a spectrum of users as possible.

Practitioners should avoid the temptation of reverting to the old familiar, less effective regulator. This may be achieved by presenting a detailed explanation of the fundamental concepts. Such an explanation should resolve any conceptual issues that may introduce doubt on the part of the practitioner.

Thus, there is a difficult compromise to be made concerning the level at which this book should be presented. It was decided that the book is intended for those beginning their studies in predictive control and for practitioners of the traditional PID. The book provides a practical introduction to the subject and, as such, does not attempt to present a full academic treatment.

1.5.2 Implementation

The concept of predictive control is very flexible and can incorporate concepts derived from other techniques. It is as effective on fast processes, such as servosystems, as it is on slow production processes such as chemical reactors. It may be implemented on a simple programmable logic controller (PLC), supervisory control system or industrial personal computer. The use of matrix calculations is not feasible on such machines and only finite difference equations are used.

One characteristic of predictive control is that the calculations are made in real time, between sampling periods, and does not require internal loop iterations (*e.g.*, "WHILE" and "FOR" loops) as is the case with multivariable control. This feature permits an industrial implementation in *block-diagram form*. Such a representation is a robust form of graphical data-flow programming that permits sequential control and continuous regulation to be used in the same controller, a requirement that is becoming increasingly important in modern control systems. Another important and practical point for the industrial practitioner using a block-diagram-based design approach is that it is quite simple to substitute one controller block for another (as long as both blocks possess the same number of inputs and outputs). The argument for the use of predictive control in block-diagram form is even more compelling when the block is simple to adjust as it brings about an immediate improvement in performance.

1.5.3 Process

Practically all industrial processes that are controllable may be regulated by predictive control. For simplicity, we will use a first-order system characterised by a gain K, a time constant T and a pure delay D. It has the advantage of being represented by a simple mathematical equation and all calculations may be carried out by hand without any difficulty. The extension to higher-order processes is immediate if the poles are real and may be equivalently represented by a series of first-order systems. In the case of complex modes, equations may also be described.

Typically, these higher-order processes are encountered in all thermal processes (*e.g.*, reactors, exchangers, ovens, air conditioning, generators); that is, the vast majority of slow industrial production processes.

1.6 Predictive Control Block Diagram

For convenience, a functional block diagram of a predictive control system is presented in Figure 1.1 including a list of commonly used notation in this work.

The objective is to regulate the process P_1 using the controller R, subject to both the measured and non-measured disturbances DV_m and DV_{nm}, respectively. In general, the disturbances DV_m and DV_{nm} manifest themselves as S_{DVm} and S_{DVnm} via the coupling interconnections P_2 and P_3, respectively. Note: controller blocks will be represented throutout this text with a clipped top left-hand corner.

The predictive controller contains the internal models M_1 and M_2 of the processes P_1 and P_2. M_1 and M_2 are identified models, possibly time dependent, implemented and used in real time in the control calculator.

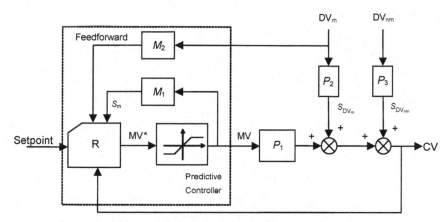

Figure 1.1. Fundamental block diagram of predictive control

The manipulated variable MV*, calculated by the regulator, is passed through a limiter producing the applied manipulated variable MV output presented to the process P_1 and to the internal model M_1 of the regulator.

The set-point at the input to the predictive regulator is denoted by Setpoint. Feedforward is a compensation signal and the variables S_m and S_{DVm} are the measured process outputs of M_1 and P_2, respectively, while the disturbance DV_{nm} and its output S_{DVnm} are not measured.

The international designations MV, CV and DV will be used to represent the manipulated variable, controlled or regulated variable and disturbances, respectively.

We are interested here in single-input, single-output (SISO) processes 1MV/1CV, which are the most commonly encountered systems in industry. However, the case of a multivariable process 2MV/2CV is also considered.

1.7 Summary

- At the beginning of the 1970s, the modelling, identification and simulation technologies had sufficiently matured.
- Digital computing had became reliable so new possibilities of industrial control appeared.
- The merging of model methodology and digital computing gave birth to model-based digital control.
- PFC was designed to replicate the model-based predictive approach instinctively employed by humans.
- The PFC controller is easy to understand, implement and tune.

2

Internal Model

Abstract. In this chapter we are mainly concerned with modelling. What kind of model to use, where to use it and how is the model implemented? A procedure to decompose the model of an integrative process is presented and the free and forced solution of linear dynamic process is reviewed.

Keywords: modelling, realigned model, independent model, disturbance, decomposition, free solution, forced solution

2.1 Why Is Prediction Necessary?

The primary concern of any financier investing in the construction of an industrial plant is to maximize the return on the financial investment. Consequently, a systems engineer will always strive to design a plant that functions on, or near, the operational boundaries dictated by the safety, technological and budgetary limitations. Ultimately, it is the operational constraints that decide the success or failure of a project. Typically, the constraints involved would arise from:

- **actuators:** *e.g.*, maximum fluid flow, heating power;
- **process limitations:** *e.g.*, maximum temperature gradient, distillation column flooding.

If "tight" process control is to be achieved then the constraints involved must be continuously respected in a dynamic fashion. As we shall see, this will lead to a requirement for some mechanism of predicting the future behaviour of the process.

Returning to the automotive example of Chapter 1, assume some incident occurs that requires the full application of the braking system. From the instant that maximum brakes are applied, the future is completely defined, *i.e.*, the stopping time and distance are fixed by factors such as the car dynamics and road conditions – both of which are out of the driver's control.

However, common sense would dictate that a more prudent approach would be to anticipate the situation and implement a controlled braking scheme commencing an appropriate distance from the potential hazard. The future prediction of the process response, *i.e.*, the stopping time and distance of the vehicle in this case, is made with the aid of a mathematical model embedded in the controller from which it derives the name "internal model".

On the other hand, if infinite deceleration (no constraint) were possible, such anticipation would not be required, as braking immediately prior to the obstacle would suffice! Thus, there is a direct causal link between "constraints" and "prediction": **Constraints → Prediction**.

In the real world, investors demand optimum profit for their investment that in turn requires optimisation of systems with inevitable constraints thus giving the overall connection between profit and prediction in the following: **Profit → Optimisation → Constraints → Prediction**.

Predictive control possesses many interesting features. However, above all, it is its natural ability to take constraints into account that led to the appearance of predictive control in the 1970s, in the oil industry in particular.

2.2 Model Types

Figure 2.1 shows a first-order process, with a gain K_p and time constant T, subject to an input e. The response of the process s_p is represented, in difference equation form, by:

$$s_p(n) = a_p s_p(n-1) + b_p K_p e(n-1), \text{ where } a_p = e^{-\frac{T_s}{T}} \text{ and } b_p = 1 - a_p,$$

where T_s is the sampling time. This may be modelled by a first-order system with a gain K_m and a time constant T_m as:

$$s_m(n) = a_m s^*(n-1) + b_m K_m e(n-1), \text{ where } a_m = e^{-\frac{T_s}{T_m}} \text{ and } b_m = 1 - a_m.$$

Note that s^* gives rise to two modelling approaches. The first approch uses a model output that is realigned with past measured or estimated process values:

$$s^*(n-1) = s_p(n-1).$$

The second approach adopts an independent process model, where a common input is supplied to both the process and model:

$$s^*(n-1) = s_m(n-1).$$

Both approaches are of use and each possesses its own individual characteristics.

2.2.1 Realigned Model

The above equations give rise to a model of the form:

$$s_m(n) = a_m s_p(n-1) + b_m K_m e(n-1),$$

where $a_m = e^{-\frac{T_s}{T_m}}$, $b_m = 1 - a_m$ and e is the input. Intuitively, we know that using the process output s_p to re-adjust the model would yield the best prediction. This operation is simplified if a difference equation form is used:

$$s_m(n-i) = s_p(n-i),$$

i.e., all past outputs of the process are measured and stored in memory. However, this representation has some limitations. If the order of the system is greater than one, and if the poles are not stable, some numerical imprecision may result and consequently the choice of sampling period becomes critical.

If a state-space representation is used, these difficulties disappear and it becomes necessary to use an "observer", *i.e.*, a mathematical procedure that reconstructs the unmeasured state variables. But, the application of this technique is not straightforward and, as a consequence, has experienced a reluctant acceptance in the production industry.

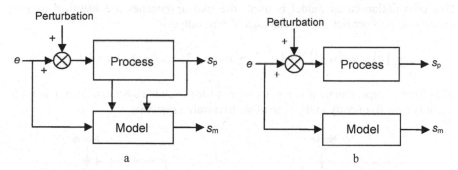

Figure 2.1. a Realigned and **b** independent models

2.2.2 Independent Model

The independent model is defined by:

$$s_m(n) = a_m s_m(n-1) + b_m k_m e(n-1).$$

In this configuration the process and independent models are supplied by the same input variable. This value is always available since it is calculated by the controller. As the process may be subjected to unknown disturbances, the resulting outputs $s_p(n)$ and $s_m(n)$ may be very different. If the disturbances are constant, the process and model outputs will evolve in parallel. Thus, the model is used to calculate a prediction of the *increment* of the process output and not to calculate the absolute response of the process subjected to a particular input.

The advantage of such an approach is that any type of model may be used, *e.g.*, mathematical, logical, or look-up table. The only requirement is that the model is capable of answering the question "What will be the resulting output increment if the process is subjected to a known input?"

Unfortunately, this approach is not valid for unstable models. In this case, using an independent model configuration as the internal model within the regulator produces an output that is open loop and unstable. However, the process may be stabilised using a stable regulator in a feedback loop. We will see later how to solve this problem.

These two approaches (realigned or independent models) result in the following identification strategies, *i.e.*, estimation of the model parameters:

- The realigned model method is equivalent to system identification using the least squares technique where the explicit variables are physical measurements derived from the process. If these variables are noisy, the identified parameters will be biased in general.
- The independent model method [2] uses only the process input and the error between the measured process and model predicted outputs, *i.e.*, the process-model error (PME). In this case the parameters of the model are generally non-biased (see Figure 2.2). Note that the distance criteria for the least-squares and model methods are usually different.

If a convolution-based model is used, the two approaches are identical as past process outputs are not used to calculate future outputs:

$$s_{\mathrm{m}}(n) = a_1 e(n-1) + a_2(n-2) + a_3 e(n-3) + \ldots + a_N e(n-N) .$$

This form of representation is not appropriate for unstable systems since N is finite but it is used frequently in the case of multivariable systems.

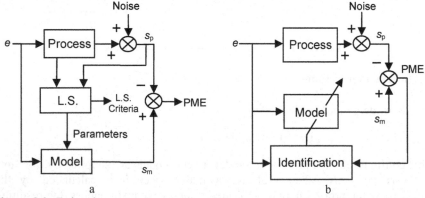

Figure 2.2. Estimation using the **a** realigned least-squares (LS) and **b** independent model methods

2.3 Decomposition of Unstable or Non-asymptotically Stable Systems

The previous comparison of the independent model concept with that of system identification gives an indication of the potential benefits that may be derived from this independent model approach, *i.e.*, absence of bias, universal application and simplicity. The case of unstable or non-minimum phase systems, *e.g.*, integrating and unstable, must be approached with caution. But first, consider the case of taking a measurable, external disturbance into account.

2.3.1 Measurable Disturbance Compensation

The real time, mathematical model implemented within the controller permits the calculation of future responses to potential manipulated variable inputs. This same technique may be used to compensate for the effects of an external, measured disturbance that is not controllable (see Figure 2.3).

Remark 2.1. In Figure 2.3, the regulated (or controlled) variable is referred to as CV and the manipulated variable is denoted by MV. These two notations will be used frequently throughout this book.

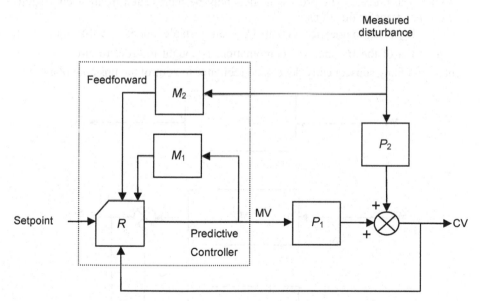

Figure 2.3. Measured disturbance compensation

The procedure consists of making future predictions of the manipulated variable and the measured disturbance using a procedure discussed in Section 2.4. An output increment is then calculated and subsequently taken into account in the control calculation (see Chapter 4).

A practical example of such compensation can be easily found. For example, assume we wish to regulate the interior temperature of a house heated by an electric radiator. The exterior temperature varies randomly but is constantly monitored. A transfer function relating the exterior and interior temperatures may be determined using system-identification techniques. This information may then be used to predict the action required to counteract the influence of the exterior temperature variations at some fixed point in the future. However, any compensatory actions must react faster than any exterior temperature variations.

This open-loop procedure should be used systematically as it does not present any risk to stability and it puts all the necessary information to best use.

2.3.2 Decomposition

For reasons of implementation and initialisation, it is necessary that the closed loop and the regulator be stable. This implies that the internal models of the regulator must also be stable. If the process is an integrating or unstable system the output, for a given steady-state, non-zero input, would escalate over time. This issue is deal with by decomposing the unstable model into two stable processes. The first model M_1 is supplied by the manipulated variable MV. The second model M_2 effectively takes the form of a compensated input and is supplied by the output of the physical process. The process models may be represented by their continuous or discrete transfer functions.

The key point to note here is that M_0 is an unstable model. A stable equivalent, in the form of the M_1 and M_2 combination, is sought in order to circumvent the inability of any subsequently developed regulator to control the unstable plant.

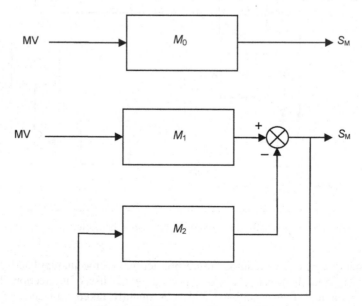

Figure 2.4. Decomposition principle

Referring to Figure 2.4: $S_M = M_1 \text{MV} - M_2 S_M$ and re-organising gives:

$S_M = \dfrac{M_1}{1+M_2} \text{MV}$. Assume we have two transfer functions, M_1 and M_2, such that

$M_0 = \dfrac{M_1}{1+M_2}$, where both M_1 and M_2 are asymptotically stable. In the nominal

case, if the process output is taken into account in the form of a compensator, the blocks may be readily identified (see Figure 2.5).

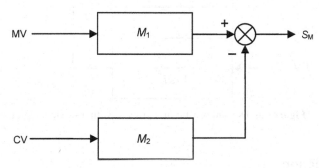

Figure 2.5. Decomposition formulation

Consider the example of a non-asymptotically stable system with an integrator represented by the continuous transfer function $H(s) = K/s$ (where s is the Laplace operator) as illustrated in Figure 2.6.

The plant is decomposed using a first-order transfer function whose dynamics resemble those of the desired closed-loop behaviour:

$$S_M = \frac{KT}{1+sT} \text{MV} - \frac{-1}{1+sT} S_M \rightarrow S_M = \frac{KT}{sT} \text{MV} = \frac{K}{s} \text{MV} \ .$$

The case where the process is unstable with an "inverse response"[2] should be dealt with cautiously (see Chapter 9). As we shall see, a satisfactory response is achieved without any particular difficulty when a coincidence point is placed beyond the characteristic 'dip' generally associated with the initial response of such systems. A typical example would be the temperature response of a chemical reactor due to the introduction of a cold reagent.

In Chapter 10 we shall see that this decomposition procedure[3] can be used beneficially in many practical situations. It should be noted that the decomposition principle is, in fact, a compromise solution between the realigned and independent model approaches.

---

[2]   A non-minimum phase system with an "unstable zero".

[3]   The benefit of this approach is that it only requires external measurements of the system model and process (input and output) and makes no assumptions about the mathematical representation, state observation or system characteristics.

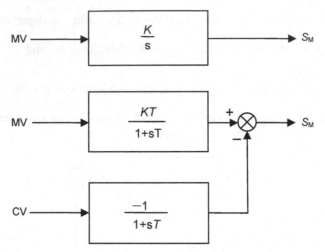

Figure 2.6. Decomposition of a process with a single integrator

2.4 Prediction

The key to any model-based controller lies in its ability to predict the process response using a model that may be physically realised, *i.e.*, a causal model. We will return to the issues of model and system identification in Chapter 7. In fact, determining the response of a process over time is equivalent to the classical problem of solving differential or difference equations. For convenience, we recall the fundamental concepts used in solving such equations.

The solution of a linear differential or finite-difference equation, from the instant $t = 0$ to the present time, consists of two terms; the free solution and the forced solution.

2.4.1 The Free Solution $S_L(t)$

The free solution (also referred to as the complementary, natural or homogeneous solution) is defined as the output when the input $e(t)$ is zero for $t > 0$, but was non-zero in the past, *i.e.*, the initial value of the process at time $t = 0$ is non-zero. This solution represents the output when no further external stimulus is applied. If the system is asymptotically stable the output will eventually decay to a zero state, as illustrated in Figure 2.7.

Example 2.1. Consider the behaviour of the interior temperature of an oven when the gas flow is cut-off. The temperature will decrease, at a rate broadly dictated by the level of insulation in the oven, towards the ambient temperature.

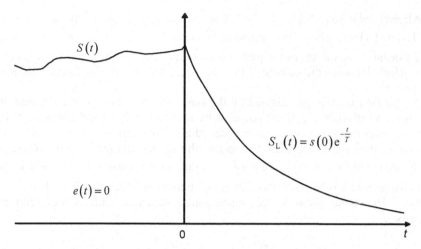

Figure 2.7. Free output response of a first-order system

2.4.2 The Forced Solution $S_F(t)$

The forced solution is also referred to as the particular, forced or inhomogeneous solution. The forced solution makes the opposite assumption to that of the free solution. This implies that all past signals, both input and output, are zero. The future, non-zero input signal is known or determined, and the future output is calculated by the model using only this non-zero input signal.

If the system is linear, the superposition theorem applies, and the future output of the process $s(t)$ is the sum of the free S_L and forced S_F responses:

$$s(t) = S_L(t) + S_F(t) .$$

Thus, the future depends both on the past process responses and its future input. The past is fixed, whereas the future input depends on the operator. Consider a first-order, linear process with a gain K and a time constant T.

Assume, for simplicity, that the future input is constant $e(t) = e_0$ (see Figure 2.8), thus,

$$S_F(t) = Ke_0 \left(1 - e^{-\frac{t}{T}}\right) .$$

The future behaviour is then the sum of the two outputs:

$$s(t) = s(0)e^{-\frac{t}{T}} + Ke_0 \left(1 - e^{-\frac{t}{T}}\right) .$$

Already we can begin to see how a simple control signal such as $MV(n) = e(n) = e_0$ allows the system to be regulated. The first term on the right-hand side takes the effects of the past into account. This, combined with the second term, which includes the effects of the regulator, determines the future output of the system.

As the first term is pre-defined by the past actions we can conclude that the only way to influence the future output is by manipulating the second term. Thus, from an intuitive standpoint, we can see that if we can somehow predict the required control signal necessary to generate the desired change in the forced term, and hence the objective $s(t)$, we have the regulator we require. We will see in the next chapter that it is beneficial to dictate the manner in which the final objective is reached. The name given to this manipulated objective path is the *reference trajectory*.

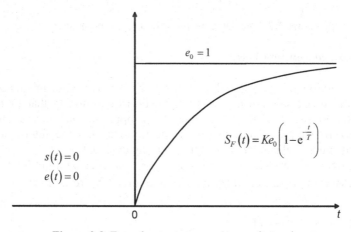

Figure 2.8. Forced output response to a unit step input

2.5 Summary

- Two generic model types are used in the formulation of PFC:

  1. **Realigned:** The input of the model is the process input and the state of the model is realigned with the measured or estimated state of the process.
  2. **Independent:** The input of the model is the input of the process and the state of the model is not realigned with the processs state. In this case the only input is the actual MV and so the model output may be different from the process output.

- Integrating processes may be represented using an independent model supplied by the applied process MV. The model is only supplied with the measured output of the process.

- Realigned and independent models are implemented in the control computer.
- The solution of a linear differential equation is composed of two terms:

    1. **Free mode:** Response of the model to a zero input;
    2. **Forced mode:** Response of the model starting from zero state and subjected to the required future input.

3

Reference Trajectory

Abstract. This chapter is devoted to the presentation and computation of the "reference trajectory" principle. A reference trajectory represents a suggested path by which the controlled variable should converge on the set-point in a specified manner. By choosing the *rate of convergence* as a controller tuning parameter we can effectively define the closed-loop time response of the system. Such a tuning parameter has the added advantage of being intuitive and instantly recognisable.

Keywords: Reference trajectory, closed-loop time response, coincidence point

3.1 Introduction

The future response of a process to a given input may be predicted using past knowledge of the process behaviour and an appropriate mathematical model. The objective of any model-based controller algorithm is to use this future prediction to modify the system output to produce the desired response.

So, "what is a desired response and how may it be specified?" The approach we adopt in answering this question is to mimic some of the natural mechanisms of human learning. Initially, this may seem unexpected, if not somewhat surprising. However, consider for a moment the underlying principles we implicitly understand from experience and that we exploit when interacting with our surroundings.

Applying a particular action to a physical process produces a specific response. For example, adjusting the controls of an electric shower mixer results, after some time delay, in a change of water temperature that will ultimately stabilise. Thus, a particular action precipitates a specific behaviour.

$$\text{Action} \rightarrow \text{Behaviour}.$$

Since we are typically interested in specific process behaviour, the question now becomes "what action is required to achieve this behaviour?" Consequently, the action is given by the inverse of the natural associative *memory* that binds action and behaviour.

Desired Behaviour → Action .

The term "behaviour" implies that it is not sufficient to assess the quality of the solution by considering the final objective value alone. The manner by which the final value is reached must also be considered, *i.e.*, the dynamic behaviour of the process that connects the initial to the final steady state value must also be taken into account. Using a geographical analogy, it is possible to take many routes from a point A to a point B, but some routes are faster and/or better than others.

Similarly, in controller design there may exist several legitimate ways to reach the objective, *i.e.*, the steady state value. It is the selected criterion or *metric* that determines the quality of the chosen solution. So, it makes sense to specify the whole future linking the present process state to its desired future state when controlling a system.

3.2 Reference Trajectory

Referring to Figure 3.1, the reference trajectory represents the temporal path we wish to follow in order to reach the desired set-point value that, for simplicity, we assume to be constant. Thus, the reference trajectory may be directly interpreted as being the desired closed-loop behaviour when the process is subjected to a set-point change. It is a time-based variable that will converge on the final set-point (be it constant or not).

It is important to realise that the reference trajectory is re-initialised *at each sampling point* using the measured or estimated process output, thus ensuring the closure of the feedback loop. At each sampling point the reference trajectory will be realigned with different trajectories, *i.e.*, the previously predicted futures. This is so because the model of the disturbed process will inevitably be imperfect and the resulting future prediction will never coincide exactly with the actual plant behaviour.

At time n the reference trajectory is initialised on the process state that converges on the set-point. The choice of function used to implement the trajectory is open-ended; it may be a look-up table, calculated analytically or may depend on the time and state of the process. Although, in general, an exponential function is used for several reasons:

- Only one point is used during initialisation, *i.e.*, the last measured or estimated process output value.
- The function is easy to calculate in real time.
- Its decrement occurs in a predictable manner, *i.e.*, the CLTR, which is taken as the time required to reach 95% of the final value and is assumed to be constant.

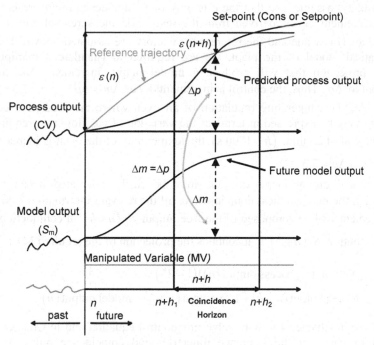

Figure 3.1. Reference trajectory

The value can vary as a function of an internal or external process variable (*e.g.*, tracking error). Usually, a future action is sought such that the predicted future response corresponds with several fixed points on the reference trajectory, which are referred to as coincidence points.

To illustrate this, an elementary case will be presented where only one coincidence point is used at $n+h$. (The choice of h will be explained in detail in Chapter 5.) Assume, without loss of generality, that the set-point is constant and that the reference trajectory is an exponential decay λ. The sampled or discrete exponential function is given by $y(n) = e^{-\frac{nT_s}{T}}$, where the decrement is $\lambda = e^{-\frac{T_s}{T}}$ and the sampling period is T_s. We are searching for the desired process output increment Δp at $n+h$. The current regulation error, *i.e.*, the difference between the set-point and the process output is $\varepsilon(n)$, and at $n+h$ the predicted error will be given by $\varepsilon(n+h) = \varepsilon(n)\lambda^h$. The desired increment Δp at $n+h$ will be given by:

$$\Delta p = \varepsilon(n+h) = \varepsilon(n) \text{ where } \Delta p = -\varepsilon(n)\lambda^h + \varepsilon(n) = \varepsilon(n)(1-\lambda^h)$$

$$\text{and } \Delta p = \big(\text{Setpoint} - \text{Process output}(n)\big)(1-\lambda^h) .$$

For simplicity, assume also that there is only one coincidence point present. The desired increment Δp is derived from the real-world measurement and the set-point value. This value and the reference trajectory are assumed known. Using a mathematical model of the plant, we are required to calculate a manipulated variable projection that will produce a model output increment Δm that is equivalent to Δp. Thus, the control is incremental, *i.e.*, $\Delta m = \Delta p$.

Consider the temperature regulation of an oven where the predictive control increment may be expressed in terms of the variation of gas flow rate required so that at the end of 20 min. $(h = 1200 \text{ s})$ the temperature of the oven is increased by 3.5 °C $(i.e., \Delta p = 3.5 \text{ °C})$.

The model output increment Δm may be easily calculated because it is evaluated in the mathematical domain where all the relevant information is known. The increment will be composed of the free output $S_L(n+h)$ in conjunction with the forced output $S_F(n+h)$ that contains the projection of the required MV:

$$\left(\text{Setpoint} - \text{process output}(n)\right)\left(1 - \lambda^h\right) =$$
$$\text{forced output}(n+h) + \text{free output}(n+h) - \text{model output}(n) .$$

We will see in Chapter 4 how to solve this control equation and in Chapter 5 the problem of choosing the reference trajectory and coincidence points will be discussed.

Generalisation. Instead of taking only one coincidence point, a coincidence horizon [*h*1 *h*2] may be used (see Figure 3.1) to minimise the quadratic error between the reference trajectory and the predicted process output on this horizon. There are numerous mathematical textbooks that describe the techniques used to minimise the relevant distance criteria, *e.g.*, minimisation of a positive-definite function in a constrained or unconstrained space.

3.3 Pure Time Delay

A predictive controller has the capacity to take any pure time delay θ associated with the process into account. We define $\theta = rT_s$, where r is the number of sample periods equivalent to the time delay, T_s is the sampling period and θ is the time delay in seconds.

Any MV applied at time n to a system with pure delay θ will have no effect on the system output until, at least, θ seconds later. In order to control such systems the reference trajectory should therefore be initialised using the predicted value of CV at sampling point $n+r$ and not the current value $CV(n)$.

We denote the *non-delayed model* as $s_m(n)$ and the process output as $s_p(n) = CV(n)$. Thus, we have:

$$s_p(n) = s_m(n-r), \text{ where } s_{\text{predict}}(n+r) = s_p(n+r) = s_m(n) .$$

We can then state that in the nominal case (model ≡ process), the increment Δ of $CV(n)$ between times n and $n+r$ is equal to the increment of the model output between times $n-r$ and n, which gives:

$$\Delta = s_p(n+r) - s_p(n) = s_m(n) - s_m(n-r),$$

and re-arranging gives:

$$s_{\text{predict}}(n+r) = s_p(n) + s_m(n) - s_m(n-r) .$$

The reference trajectory is initialised using $s_{\text{predict}}(n+r)$ and not the measured value $s_p(n)$. Note that an actual "future prediction" is not made; it is simply an extrapolation, in time, of the current value using the model's response. In effect, the "future" outputs until time $n+r$ have already been determined and may not be further influenced by any inputs occurring after time n. Also note that this assumption is valid only when the model and all the inputs acting on the process are known.

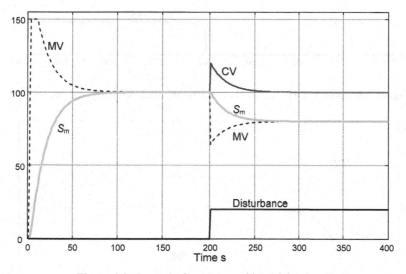

Figure 3.2. Control of a process with no delay ($r = 0$)

Proceeding in this manner eliminates the effects of the pure delay. Consider Figures 3.2 and 3.4. The behaviour of the MV acting on the process without time delay is identical to that of the process with time delay for a change in set-point with no disturbance present. However, there is a shift of r time samples in the response of the CV in the delayed case.

To illustrate this, consider the nominal case of a first-order process defined by $K = 1$, $T = 30$ s, $r = 20$ s, $T_s = 1$ s, CLTR = 50 s, with a constraint $MV_{max} = 150$ and is subject to an additive disturbance of +20 on the output at 200 s.

Figures 3.3 and 3.4 show the behaviour for a change in set-point. The figures also demonstrate the effects of a disturbance acting on the output of two identical systems, one with a time delay and the other with no time delay.

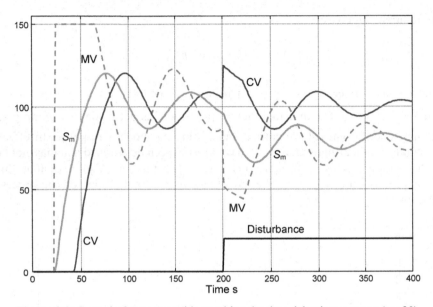

Figure 3.3. Control of a process without taking the time delay into account ($r = 20$)

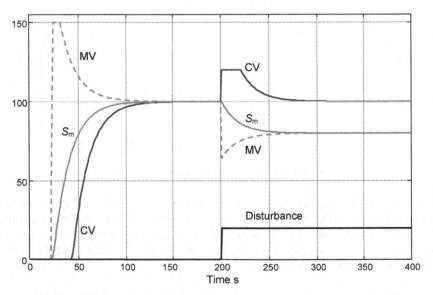

Figure 3.4. Control of a process taking the time delay into account ($r = 20$)

3.4 Summary

- The path to the set-point, constant or time varying, is defined by a time-based reference trajectory that links the actual position of the CV to the final set-point.
- The selection of the reference trajectory is open. But, for simplicity sake, an exponential function is usually adopted.
- The decrement of the exponential function determines the closed-loop time response.
- On this trajectory a certain number of coincidence points are selected where the future CV response should coincide with the reference trajectory.

4

Control Computation

Abstract. In this chapter a procedure for computing the MV associated with the desired future CV projection for different types of processes is presented. The benefits of projecting the future $MV(n+i)$ onto a set of basis functions, as opposed to computing the whole vector of the future MV, are demonstrated. This approach reduces the question of predictive control to determining the *coefficients* of these MV polynomials. The concepts of feedforward and time-delay compensation are examined and a procedure for dealing with complex poles is also presented. Finally, the issues of implementing PFC in practice and the bumpless transfer from PID control to PFC are discussed.

Keywords: Basis function, convolution representation, feedforward, time delay, complex poles, decomposition principle, bumpless transfer

4.1 Elementary Calculation

We have seen that the introduction of the relatively simple concept of a reference trajectory ensures that the desired process output increment is reached at a future point $n+h$ (see Section 3.2). In its basic form, specifying a single coincidence point is all that is required to define the predictive controller. The real-world process increment Δp is then transferred to the mathematical domain where we have knowledge of the model and its state variables via the equation $\Delta m = \Delta p$, where Δm is the equivalent model increment.

One interpretation of this approach would be to assume that we have the ability to slow down, or stop, time. It would then be feasible to test potential scenarios using some form of process simulation. The future manipulated variable projection is selected that best approximates the desired increment while tracking the reference trajectory as closely as possible.

The simplest selection technique would be to search for some future $MV(n+i)$ that would result in a predicted output increment Δm of the process simulation equivalent to the desired physical process Δp at a future time $n+h$.

The same procedure would then be repeated at the next sampling point. There are two fundamental issues that must be considered:

1. How to determine future model outputs.
2. How to structure the future MVs.

4.1.1 How to Determine Future Model Outputs

We saw in Section 2.4 that the future output depends on past the MVs that drive the process model into its current state. The absence of any future control signal would give rise to a non-zero output, referred to as the *homogeneous, complementary* or *free* output, which is a natural consequence of the system dynamics. The future model output also depends on the effects of any current and future inputs $MV(n+i)$ and it is these values that are sought. Thus, the output consists of two terms:

- Effects of the previous inputs that are known but cannot be altered, *i.e.*, the influence of the past.
- The effects of future inputs that are open to manipulation, *i.e.*, the forced output that depends on the future MV.

4.1.2 How to Structure the Future MVs

Many techniques have been proposed to impose structure on the future values of the MV. If we assume that the coincidence point is placed at time $n+h$, it would be necessary to calculate $h-1$ future values of the MV.

The most intuitive approach is to assume that no restrictions apply and to calculate, at every moment, all the future $MV(n+i)$ values. But this does not represent the best solution, for reasons that will be discussed in Section 4.5.

Another approach would be to restrict the possible values of $MV(n+i)$ that may be selected. This reduced set of potential alternatives is then searched to find the $MV(n+i)$ projection that best approximates the desired, predicted output at the coincidence point.

The most elementary case is given by $MV(n+i) = MV(n)$ where $0 = i \le h-1$, *i.e.*, a constant future MV. The procedure is re-calculated at each sampling point. In this case, it is the known forced output response of the internal model to a unit step that is multiplied by the $MV(n)$ value. This ensures that the model output coincides with the reference trajectory at $h-1$.

Consider the simple case of a first-order system sampled at a rate of T_s seconds, with a gain K_p and decrement a_p. The output of the process $\left(s_p(n) = CV(n)\right)$ is given by:

$$s_p(n) = s_p(n-1)a_p + (1-a_p)K_p MV(n-1),$$

where $a_p = e^{-\frac{T_s}{T}}$ and the model output is given by:

$$s_m(n) = s_m(n-1)a_m + (1-a_m)K_m MV(n-1),$$

and $a_m = e^{-\frac{T_s}{T_m}}$. As was seen in Chapter 3, the increment Δp may be expressed as:

$\Delta p = \left(\text{Setpoint} - s_p(n)\right)l_h$, assuming a constant set-point, with $l_h = 1 - \lambda^h$,

where $\lambda = e^{-\frac{3T_s}{T_r}}$, with $T_r = \dfrac{\text{CLTR}}{3}$ and h representing the coincidence point. For a first-order system, the 95% response time ▬ $3 \times$ (system time constant), approximately. What is the value of Δm?

$$\Delta m = \text{free output}(n+h) + \text{forced output}(n+h) - s_m(n),$$

where free output$(n+h) = s_m(n)a_m^h$, forced output$(n+h) = MV(n)K_m(1-a_m^h)$.
The free mode is obtained by observing that:

$$s_m(n+1) = s_m(n)a_m + (1+a_m)K_m.0,$$

$$s_m(n+2) = s_m(n+1)a_m = s_m(n)a_m^2,$$

$$s_m(n+3) = s_m(n+2)a_m = s_m(n)a_m^3,$$

$$\ldots$$

$$\text{free output}(n+h) = s_m(n)a_m^h.$$

The forced mode, which we are seeking, is the response of the process to a step input of unknown amplitude $MV(n)$. $MV(n)$ is obtained by the combination of:

$$s_m(n+1) = s_m(n)a_m + (1-a_m)K_m MV(n),$$

with the initial condition $s_m(n) = 0$.

Since $MV(n) = MV(n+1) = MV(n+2) = \ldots = MV(n+h)$, we obtain:

$$\text{forced output}(n+h) = MV(n)K_m(1-a_m^h).$$

Thus, the model output increment is given by:

$$\Delta m = s_{\mathrm{m}}(n)a_{\mathrm{m}}^{h} + \mathrm{MV}(n)K_{\mathrm{m}}\left(1-a_{\mathrm{m}}^{h}\right) - s_{\mathrm{m}}(n) \ .$$

The target $\Delta p = \Delta m$ is fulfilled by:

$$\left(\mathrm{Setpoint} - s_{\mathrm{p}}(n)\right)l_{h} = s_{\mathrm{m}}(n)a_{\mathrm{m}}^{h} + \mathrm{MV}(n)K_{\mathrm{m}}\left(1-a_{\mathrm{m}}^{h}\right) - s_{\mathrm{m}}(n) \ .$$

Extracting the manipulated variable $\mathrm{MV}(n)$ gives:

$$\mathrm{MV}(n) = \frac{\left(\mathrm{Setpoint} - s_{\mathrm{p}}(n)\right)\left(1-\lambda^{h}\right) - s_{\mathrm{m}}(n)a_{\mathrm{m}}^{h} + s_{\mathrm{m}}(n)}{K_{\mathrm{m}}\left(1-a_{\mathrm{m}}^{h}\right)} \ .$$

If we set $b_{\mathrm{mh}} = 1 - a_{\mathrm{m}}^{h}$ and $l_{h} = 1 - \lambda^{h}$, then we get:

$$\mathrm{MV}(n) = \frac{\left(\mathrm{Setpoint} - s_{\mathrm{p}}(n)\right)l_{h} + s_{\mathrm{m}}(n)b_{\mathrm{mh}}}{K_{\mathrm{m}}b_{\mathrm{mh}}} \ .$$

This equation represents the control law in its most basic form. However, this form is adequate for many applications and it is commonly encountered in industrial control systems. Many extensions are possible: several coincidence points, various structures of the MV, the use of reference trajectories other than an exponential, *etc.* Regardless of the variant used, the equation will always possess the same form, *i.e.*,

$$\mathrm{MV}(n) = \frac{\mathrm{Desired\ increment} - \mathrm{Free\ Output\ increment}}{\mathrm{Unit\ forced\ output}} \ .$$

4.2 No Integrator?

In practice, this simple equation will be embedded in the regulator in the form:

$$\mathrm{MV}(n) = k_{0}\varepsilon(n) + k_{1}s_{\mathrm{m}}(n), \ \mathrm{where} \ k_{0} = \frac{\left(1-\lambda^{h}\right)}{K_{\mathrm{m}}\left(1-a_{\mathrm{m}}^{h}\right)} \ \mathrm{and} \ k_{1} = \frac{1}{K_{\mathrm{m}}} \ .$$

The equation consists of a proportional error term to which is added the normalised model output. It is worth noting that there is no explicit integrator present in this equation – this may appear strange to PID users.

The primary function of any process controller is to remain stable under all operating conditions and generate a zero positional error in the presence of constant disturbances. Regulators should also reject the effects of any structural or process variations that may occur subsequently. In the standard PID regulator, it is the purpose of the integrator term to meet these requirements.

In steady state, assume that we have a stable regulator with a gain mismatch between the internal model and the physical process, $i.e., K_p \neq K_m$, as well as a constant additive disturbance on the CV. A question that may be posed is what point is $\text{Setpoint} - \text{CV}(n) = 0$? Gathering the MV terms in the control equation, we get:

$$\left(\text{MV}(n) K_m - s_m(n)\right)\left(1 - a_m^h\right) = \left(\text{Setpoint} - \text{CV}(n)\right)\left(1 - \lambda^h\right),$$

where $\left(1 - \lambda^h\right)$ and $\left(1 - a_m^h\right)$ are finite and non-zero. (If we assume that $\text{Setpoint} - \text{CV}(n) = 0$, this also implies that $\text{MV}(n) K_m - s_m(n) = 0$.)

Thus, assuming a stable steady state regulator exists, the above control expression will always be valid as the equation is independent of the process and of any disturbances.

Note that zero positional error is achieved without the presence of an explicit integrator. The merits of this approach become clear when all the difficulties that are associated with the programming and management of the integrator in the PID regulator are taken into consideration.

In light of the above result, the configuration of the regulator's internal model needs to be re-visited. In Chapter 1 we saw that a choice existed between the *independent* model given by:

$$s_m(n) = s_m(n-1) a_m + \left(1 - a_m\right) K_m \text{MV}(n-1).$$

The *realigned* model, which mixed the model's parameters with those of the process state variables, is given by:

$$s_m(n) = \text{CV}(n-1) a_m + \left(1 - a_m\right) K_m \text{MV}(n-1).$$

If the same equation is now reconsidered using the realigned model approach, we find that $K_m \text{MV}(n) - \text{CV}(n) = 0$ is never satisfied. Thus, we find that the intuitive approach of realigning the model with the process at each sampling time is not satisfactory. Nevertheless, we will see in Chapter 8 that such a configuration does possess some benefits.

Therefore, we note that if the process loop is stable and the set-point and disturbances are constant then no bias appears in the steady state, *i.e.*, no permanent error. The presence of an integrator in a PID controller is crucial in preventing steady state offset. But, there are practical drawbacks associated with the initialisation and integral saturation (integral wind-up); with PFC, all these problems disappear.

4.3 Basis Functions

There are three reasons why the future MV should be structured around a *basis function*:

- nature of the MV;
- calculation complexity;
- nature of the control.

Each will be discussed in the following sections. In general, the vector of future MVs is not established directly. Instead, we determine the *projection* μ_j of the MV onto a finite set of basis functions:

$$F_j(i) \rightarrow \text{MV}(n+i) = \sum \mu_j F_j(i),$$

where $j = 0, \ldots, N-1$ and $0 \le i \le h$. Thus, the MV is expressed as a weighted sum of a finite number N basis functions.

The $F_j(i)$ functions may be known or tabulated. $G_j(i)$ is the vector of corresponding outputs when $F_j(i)$ is applied to the internal model of the regulator with zero initial conditions. In particular, the case where each basis function consists of a polynomial basis, *i.e.*, $F_j(i) = i^j$ is referred to as *Predictive Functional Control*. In the elementary case described above, the basis functions reduce to $N = 1$, $F_0(i) = i^0 = 1$.

4.3.1 What MVs Are Required?

Generally, the physical processes required to be controlled are low-pass in nature. Precautions must be taken when calculating the MV to prevent the serious risk of adding a high-frequency parasitic to the useful MV. The level of resulting parasitic is dependent on the particular method used to solve the control equation, *i.e.*, the solver.

A typical example of such a condition is illustrated in Figure 4.1. In this case, the behaviour of the regulated output CV is correct for the noise-free case (see Figure 4.2) but the resulting MV is unacceptable. In fact, industrial specifications are not only concerned with the quality of the CV, but also deal with the "smoothness" of the MV, for reasons such as actuator fatigue, energy conservation, *etc*.

Two strategies are suggested for controlling the MV and each requires a good appreciation of the following issue: The control calculator equation achieves its goal by minimising the distance between the reference trajectory and the predicted output [8, 12, 14] using some criteria or *metric*, which is often quadratic. The components of the calculator equation may vary, depending on the type of control behaviour required.

The first strategy introduces a weighting coefficient μ to the control equation, which acts as a tuning parameter. This dictates the contribution of the level of derivative activity of the MV. But what value should be given to this weighting coefficient?

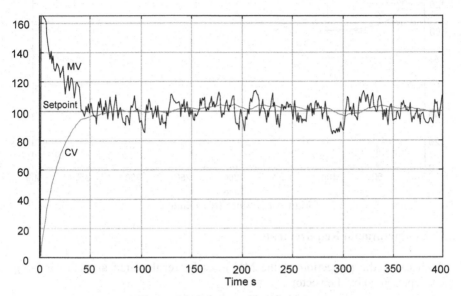

Figure 4.1. Solution with noise present

Assume that we have to control the temperature of a vessel via the flow of a heating fluid. In this case the valve opening and the temperature represent the MV and CV, respectively. Thus, the dimension of the tuning parameter μ would be $(\text{temperature/speed of valve opening})^2$. There is no systematic procedure, apart from trial and error, for selecting the value of μ for the local person tuning the controller.

This weighting term influences the dynamics of the CV in a manner that is not obvious. The ideal solution would involve choosing the best control possible without applying any restrictions on the projection of the MV. However, some solutions might not be feasible because of physical or economical constraints. Consequently, a small, practical, corrective term is introduced into the final solution *a posteriori* to deal with these restrictions.

The second strategy involves limiting, *a priori*, the nature of the MV projection to a well-defined set of basis functions. Figure 4.2 shows a low-pass filtered process controlled using strategy 1. Figure 4.1 shows the same process controlled using the second strategy. Both regulators produce CVs that are similar in nature but are generated from very different MVs. When the preceding discussion is taken into consideration, it is clear that the MV in Figure 4.1 would not be acceptable in an industrial context.

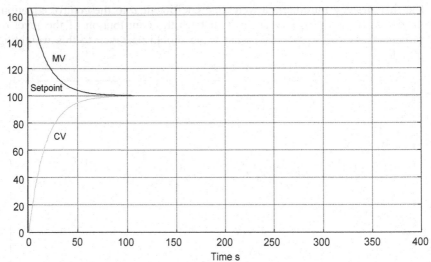

Figure 4.2. Noise-free solution

4.3.2 Computational Requirements

In both cases, the projection of the MV must be recalculated at each sampling point, represented by the vector

$$\left[MV(n+i) \right] = \left[MV(n) \, MV(n+1) \, ... \, MV(n+h-1) \right] .$$

This ensures that the set-point is achieved at the desired coincidence point $n+h$. We will see in Chapter 6 that the complete vector will be used to respect the state constraints, although, in most cases, only the *first* value is utilised, *i.e.*, $MV(n)$. At the next sampling point, the complete $MV(n+i)$ vector changes and will differ from the previous vector. In order to ensure the stability of the regulator, it is necessary to calculate h values of the MV even though only one MV value is applied. It is quite possible that h might be large. Consequently, the extent of the calculations required might give rise to implementation issues on devices such as programmable logic controllers (PLCs).

4.3.3 Polynomial Set-point

Typical applications, such as a mechanical servomechanism or a chemical reactor, have variable set-points that may often be approximated by a relatively low-order polynomial, *e.g.*, less than 4.

Assume that the process is asymptotically stable in open loop (integrator free) with a ramp set-point. In steady state, it is necessary and sufficient that the MV is a ramp[4] to guarantee that the CV is a ramp. It seems natural to search for an MV in the form of:

$$MV(n+i) = \mu_0 F_0(i) + \mu_1 F_1(i) = \mu_0.1 + \mu_1 i, \text{ each } \mu \text{ is unknown.}$$

Thus, we have two unknowns μ_0 and μ_1. In its simplest form two equations may be used to determine the two coincidence points h_1 and h_2.

Every "input basis function" $F_j(i)$ generates a unique "output basis function" $G_j(i)$, tabulated *a priori* or calculated locally. In this case, the step basis function input results in a step and an exponential basis function output. Similarly, the ramp input basis function produces a response composed of a ramp and an exponential in the transient phase. By linear superposition the output is the sum of the individual elementary outputs. This results in a forced output given by:

$$\mu_0 G_0(i) + \mu_1 G_1(i) \text{ for an input } MV(n+i) = \mu_0 F_0(i) + \mu_1 F_1(i).$$

For coincidence points h_1 and h_2, the corresponding desired increments are denoted by $\Delta p(h_1)$ and $\Delta p(h_2)$. The resulting system of equations is given by:

$$\Delta p(h_1) = \text{Free output}(n+h_1) + \mu_0 G_0(h_1) + \mu_1 G_1(h_1) - s_m(h_1),$$

and

$$\Delta p(h_2) = \text{Free output}(n+h_2) + \mu_0 G_0(h_2) + \mu_1 G_1(h_2) - s_m(h_2),$$

from which we may extract the optimal values $\mu_{0_{opt}}$ and $\mu_{1_{opt}}$ by some method, assuming that a solution exists. The resulting $MV(n)$ is given by:

$$MV(n) = \mu_{0_{opt}} F_0(0) + \mu_{1_{opt}} F_1(0).$$

However, for $F_1(0) = 0$ and $G_1(0) = 0$, $MV(n)$ reduces to

$$MV(n) = \mu_{0_{opt}} F_0(0) = \mu_{0_{opt}} \text{ since } F_0(0) = 1.$$

---

[4] This is a property of eigenfunctions of linear systems when the input and output are of the same nature.

It should be noted that when the basis functions are reduced to a step only $\mu_{0_{opt}}$ is not equal to $MV(n)$.

Consider the theoretical example of a first-order process with $K_p = K_m = 1$, a time constant $T = 935$ s and a pure time delay of $r_d = 10$ s. The desired set-point is a ramp with a slope of $s_{lp} = \dfrac{0.5}{T_s}$ where the sampling time is $T_s = 3$ s. The desired process error in this case is:

$$\Delta p = \text{Setpoint}(n+h) - \text{Setpoint}(n) + \left[\text{Setpoint}(n) - CV(n)\right]l_h .$$

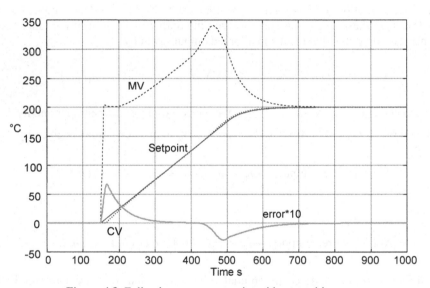

Figure 4.3. Following a ramp set-point without tracking error

The objective is to reach a constant set-point of value Setpoint $= 200$ at $n = 500$. Step and ramp basis functions are chosen with $h_1 = 50$, $h_2 = 100$ and CLTR $= 300$ s. Notice that the error, given by $\varepsilon(n) = \text{Setpoint}(n) - CV(n)$, is tracked and also that the set-point is followed with zero lag after the transient phase. If a parabolic set-point is chosen, a small tracking error appears because the parabolic set-point does not map itself entirely onto the basis functions (see Figure 4.3). The three benefits associated with choosing the MV from basis functions are:

- better performance;
- simpler calculations;
- ease of implementation.

It is also worth noting that the regulator in the previous example does not explicitly possess two integrators.

4.4 Extension

If the process contains M integrators, and assuming we have an Nth-order polynomial controller, then it is possible to reduce the highest order of the basis function to $N - M$. In fact, using a higher-order controller has no influence on its tracking ability.

However, it is always necessary to choose at least one with similar characteristics to the step basis function as it permits a non-zero MV at the instant n:

$$\left(\forall_{\substack{F_j (0) = 0 \\ (j \neq 0)}} \right).$$

It is quite easy to see the effects of designing a controller without the use of a non-zero basis function, such as a step. The controller apparently satisfies the predicted, steady state output at the coincidence points h_1 and h_2 with an $\mathrm{MV}(n) = 0$. It may be seen that the MV does not depart from zero because there is no basis function with the capacity to initiate the growth at n.

The detailed choice of coincidence points will be addressed in Chapter 5. But, for now, an empirical rule of $h_2 = \dfrac{\mathrm{CLTR}}{3}$ and $h_1 = \dfrac{\mathrm{CLTR}}{6}$ may be applied in the case of a ramp set-point.

4.5 Implicit Regulator Calculation

Consider the elementary case of a first-order system. The internal model may be given by:

$$s_m (n) = s_m (n-1) a_m + (1 - a_m) K_m \mathrm{MV}(n-1) .$$

The regulator equation is given by:

$$\mathrm{MV}(n) = \frac{\left(\mathrm{Setpoint} - s_p (n) \right)\left(1 - \lambda^h \right) - s_m (n) a_m^h + s_m (n)}{K_m \left(1 - a_m^h \right)} .$$

An equivalent transfer function for the regulator may be determined by substituting for the model output s_m in the regulator equation and solving for MV in terms of the regulation error $\varepsilon(n)$. Given that $\varepsilon(n) = \mathrm{Setpoint} - s_p (n)$, we have:

$$\mathrm{MV}(n) = \varepsilon(n) Q(h) + \frac{s_m (n)}{K_m}, \quad \text{where } Q(h) = \frac{1 - \lambda^h}{K_m \left(1 - a_m^h \right)} .$$

Taking the z transform of the equations $s_m(n)$ and $MV(n)$ gives:

$$s_m(z) = K_m \frac{(1-a_m)z^{-1}}{1-a_m z^{-1}} MV(z),$$

and

$$MV(z) = \varepsilon(z)Q(h) + \frac{s_m(z)}{K_m} .$$

Substituting for $s_m(z)$ gives:

$$MV(z) - MV(z)\frac{(1-a_m)z^{-1}}{(1-a_m z^{-1})} = \varepsilon(z)Q(h),$$

where $MV(z)\dfrac{(1-z^{-1})}{1-a_m z^{-1}} = \varepsilon(z)Q(h)$.

Thus, $R(z)$ is an equivalent regulator transfer function given by:

$$R(z) = \frac{MV(z)}{\varepsilon(z)} = Q(h)\frac{(1-a_m z^{-1})}{1-z^{-1}} .$$

The time-domain equivalent of $MV(z) = Q(h)\dfrac{(1-a_m z^{-1})}{(1-z^{-1})}\varepsilon(z)$ is given by:

$$MV(n) = MV(n-1) + Q(h)\left[\varepsilon(n) - a_m \varepsilon(n-1)\right] .$$

This equation represents an integrator-based controller, which appears explicitly in this formulation; whereas it is contained implicitly in the control equation form. Re-organising, it may be seen that this controller is merely a PI regulator, *i.e.*,

$$MV(z) = k_{prop}\varepsilon(z) + \frac{k_{intg}}{1-z^{-1}}\varepsilon(z) .$$

The equivalent Laplace regulator is given by:

$$MV(s) = k_p \varepsilon(s) + \frac{1}{sT_i}\varepsilon(s),$$

where k_p is the proportional gain and T_i is the integral action time constant.

A valid question may be posed at this point; namely, Is there any benefit in applying a predictive regulator to a first-order process if an equivalent PI regulator may be easily found?

There is no particular benefit to be gained from applying a predictive controller if the PI tuning is straightforward. An example of such a system would include a supply line, flow control with level 0 regulation flow indicator controller (F.I.C). However, the absence of an explicit integrator is a significant issue. If the process contains a pure time delay, the classical Smith predictor compensator may be applied. But this also requires the use of a model-based approach method. In this case, the benefit of using predictive control is derived from the regulator's ease of tuning.

If the process contains MV constraints, CV constraints, has a high order or is time varying, then the use of a PID or Smith predictor is not a viable option. In such circumstances a predictive controller must be used.

Nevertheless, it is reassuring, from a theoretical perspective, to find a problem where there is equivalence among methods and, furthermore, a freedom of choice for the practitioner. The intention here is to demonstrate another alternative for the practitioner's repertoire and not to find a replacement for PID.

4.6 Control of an Integrator Process

Consider an integrator process system with a first-order transfer function and pure time delay. A typical industrial example would be level control where the dynamics of the transportation lag, sensor and actuator are taken into account. An example of such a process would be:

$$CV(s) = MV(s)\frac{K_p e^{-sT_r}}{s(1+sT)} = MV(s)M_0(s) .$$

In practice, a PID regulator would have difficulty in controlling such a process. As already discussed in Chapter 3, it is standard practice to deal with a pure time delay by time shifting the reference trajectory, which is initialised by:

$$CV_{pred}(n) = CV(n) + s_m(n) - s_m(n-r),$$

where $r = \text{round}\left(\dfrac{T_r}{T_s}\right)$, $T_r = $ time delay (in s) and $T_s = 1$ s.

$s_m(n)$ represents the model output without time delay and is given by:

$$s_m(n) = s_m(n-1)a_m + Kb_m CV(n-1), \text{ where } a_m = e^{-\frac{T_s}{T}} \text{ and } b_m = 1 - a_m .$$

The process may be modelled without time delay in difference equation form by:

$$s_m(n) = s_m(n-1)a_m + Kb_m CV(n-1) .$$

The integrator is decomposed using the procedure illustrated in Figure 2.6. The results of its application are shown in Figure 4.4. The process M_1 in Figure 2.4 is replaced by two, first-order processes in parallel:

$$M_{11} = \frac{K_1}{1+sT} \quad \text{and} \quad M_{12} = \frac{K_2}{1+sT_{dec}}, \text{ where } T_{dec} \text{ is a decomposition time}$$

constant in the region of $\dfrac{\text{CLTR}}{3}$ with $K_2 = \dfrac{K_p T_{dec}^2}{T_{dec} - T}$, $K_1 = K_p T_{dec} - K_2$, $a_s = e^{-\frac{T_s}{T_{dec}}}$

and $b_s = 1 - a_s$.

The return path M_2 is taken as $M_2 = \dfrac{-1}{1+sT_{dec}}$. It may be verified that M_0 is

given by $M_0 = \dfrac{M_1}{1+M_2}$. The forward (or direct) path $s_d(n)$ under these conditions

is:

$$s_d(n) = s_{m1}(n) + s_{m2}(n) \text{ with } s_{m1}(n) = s_{m1}(n-1)a_m + b_m K_1 e(n-1),$$

and

$$s_{m2}(n) = s_{m2}(n-1)a_s + b_s K_2 e(n-1) .$$

The feedback branch is given by:

$$s_{mr}(n) = s_{mr}(n-1)a_s + b_s CV(n-1) .$$

Thus, the complete decomposed output model $s_m(n)$ is given by:

$$s_m(n) = s_{m1}(n) + s_{m2}(n) + s_{mr}(n) .$$

The incremental contribution SS of the regulated variable, which makes a zero-order or *flat* prediction, i.e., $s(n+i) = s(n)$, is composed of an incremental free term and a forced output term (see Section 4.7).

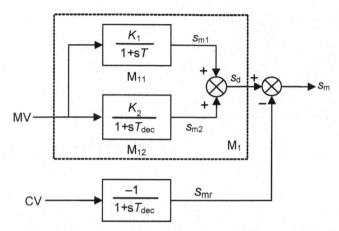

Figure 4.4. Decomposition of a first-order process with integrator

Assuming a coincidence point of h and substituting $b_{mh} = 1 - a_m^h$, $b_{sh} = 1 - a_s^h$ and $l_h = 1 - e^{-\frac{3hT_s}{TRBF}}$ we get: $SS = s_{mr}(n)b_{sh} - CV_{pred}(n)b_{sh}$. Then, the control equation is given by:

$$MV(n) = \frac{(\text{Setpoint} - CV(n))l_h + s_{m1}(n)b_{mh} + s_{m2}(n)b_{sh} - SS}{K_1 b_{mh} + K_2 b_{sh}}.$$

In Figure 4.5, we have:

$$K_p = 0.05, \ T = 30 \text{ s}, \ T_r = 40 \text{ s}, \ T_s = 1 \text{ s}, \ h_s = 100, \ CLTR = 200 \text{ s}.$$

Remark 4.1. We will see later in Chapter 7 that an alternative procedure called transparent control may also be used. The identification of an open-loop integrative process is always risky. Consequently, a generally accepted procedure, used by operators, involves firstly closing the loop using a P controller only with a manually tuned gain to stabilise the process and then identifying the closed-loop process. A PFC regulator is then added to achieve the required dynamic performance.

4.7 Feedforward Compensation

Measured disturbance compensation is a procedure of great practical benefit, which should be incorporated as often as possible into the regulator design process. This procedure was introduced in Chapter 2. The underlying concept is simple: to counteract the effects of a disturbance before it appears.

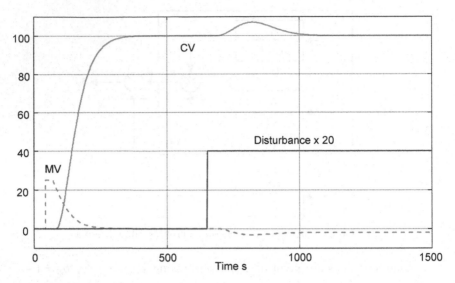

Figure 4.5. Control of an integrator process with time delay and time constant

4.7.1 Process Without Time Delay

The question that needs to be addressed at this point is whether or not it is possible to act on a process *a priori* such that the action produced opposes the effects of a disturbance that will occur at some point in the future?

It is important to realise that the dynamics of the corrective action must be faster than those of the disturbance, *i.e.*, without waiting for the feedback loop to generate an error signal that could act on the disturbance too late.

The disturbance at instant $n+h$ creates a control increment Δ_{pert} that depends on the free and forced outputs of the process. The past measured disturbance produces an output $s_{m_{pert}}(n)$ in response to the known transfer function P_2 (see Figure 4.6). Under these conditions, the free output $S_{L_{pert}}(n+h)$, which depends only on the past measured disturbance, is known. On the other hand, the forced output $S_{F_{pert}}(n+i)$ is unknown and therefore a prediction of the disturbance must be made. The simplest extrapolation of $\text{Pert}(n)$ is a "flat" extrapolation of order zero, *i.e.*,

$$\text{Pert}(n+i) = \text{Pert}(n), \ 0 < i < h-1 \ .$$

This results in a step characteristic response of the process multiplied by the local value of the measured disturbance:

$$S_{F_{pert}}(n+h) = G_0(h)\text{Pert}(n), \text{ where } G_0(h) \text{ is a gain function of } h \ .$$

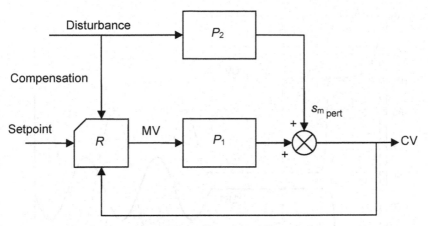

Figure 4.6. Feedforward configuration

A higher-order extrapolation may also be made on the past horizon and used to identify the correct $(q_0,\ q_1)$ reflecting the past values of $\text{Pert}(n-i)$ more closely,

$$i.e.,\ \text{Pert}(n+i) = q_{0_{\text{opt}}} + q_{1_{\text{opt}}} i \ .$$

Alternatives include the use of a higher-order polynomial extrapolation, or an extrapolation based on known basis functions such as sinusoids. A perturbation model consisting of:

$$s_{m_{\text{pert}}}(\text{output}),\ S_{F_{\text{pert}}}(\text{forced output})\ \text{and}\ S_{L_{\text{pert}}}(\text{free output})$$

is then introduced and the term $\Delta\text{pert}(n+h)$ is added to the control equation, giving:

$$\left(\text{Setpoint} - s_p(n)\right)l_h = s_m(n)a_m^h + \text{MV}(n)k_m\left(1 - a_m^h\right) - s_m(n) + \Delta\text{pert}(n+h)$$

where $\Delta\text{pert}(n+h) = S_{L_{\text{pert}}}(n+h) + S_{F_{\text{pert}}}(n+h) - s_{m_{\text{pert}}}(n)$,

from which $\text{MV}(n)$ may be extracted:

$$\text{MV}(n) = \frac{\left(\text{Setpoint} - s_p(n)\right)l_h + s_m(n) - \left(s_m(n)a_m^h + S_{L_{\text{pert}}}(n+h) + S_{F_{\text{pert}}}(n+h) - s_{m_{\text{pert}}}(n)\right)}{K_m\left(1 - a_m^h\right)} \ .$$

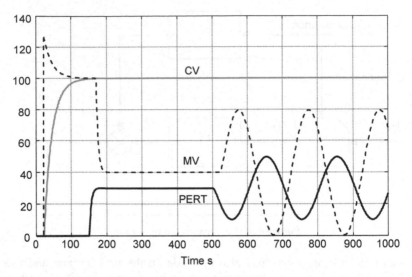

Figure 4.7. Control with feedforward disturbance compensation

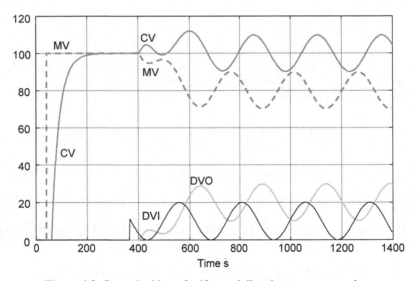

Figure 4.8. Control without feedforward disturbance compensation

All modelled, measurable disturbances may be taken into account in an additive fashion since the process is linear. Consider the simple example of a first-order process P_1 with a gain $K_1 = 1$ and a time constant $T_1 = 30$ s. There is also a disturbance process with gain $K_2 = 2$, time constant $T_2 = 50$ s, $h = 1$ and CLTR = 70 s. The disturbance consists of a step filtered with a time constant $T_3 = 5$ s with a sinusoid added at 500 s. The results, with and without compensation, are shown in Figures 4.7 and 4.8.

4.7.2 Process with Time Delay

Consider the case where the delay r_p affecting the driving process P_1 is smaller than that of the delay r_{pert} affecting the process P_2. In general, this type of compensation is only effective in applications where the dynamics of P_1 are faster than those of the process P_2. In cases where the disturbances are low frequency relative to those of the process dynamics, improvement by the usual feedback loop is satisfactory and the benefits resulting from disturbance compensation are minimal. Thus, it is advisable to specify clearly the operating conditions in which the procedure is of practical use, not because it disturbs the operation of the process but because it is of little practical benefit to the regulator's operation!

Remark 4.2. An analysis of the reverse situation $\left(r_p > r_{pert} \right)$ is also possible. But, such a study would prove of little interest since the MV signal arrives after that of the disturbance with the result that its effect would be minimal. The implementation of such a scheme would also prove more complex.

The process delay r_p is taken into account in the usual way; namely, a compensation term is added to the control equation that incorporates the delay differences $d_r = r_{pert} - r_p$ affecting the forced output. Assume the disturbance model is of the form:

$$s_{m_{pert}}(n) = s_{m_{pert}}(n-1)a_p + k_p\left(1-a_p\right)\mathrm{pert}\left(n-r_{pert}\right).$$

The time delay of compensation term is given by:

$$S_{pred} = s_p(n) + s_m(n) - s_m(n-r_p).$$

Thus, the process transfer function becomes:

$$S_{pert} = s_{m_{pert}}(n)\left(1-a_p^h\right) - \mathrm{pert}\left(n-d_r\right)k_p\left(1-a_p^h\right),$$

and the disturbance model:

$$\left(\mathrm{Setpoint} - S_{pred}\right)l_h = s_m(n)a^h + \mathrm{MV}(n)K_m\left(1-a^h\right) - s_m(n) + S_{pert}.$$

In the case of Figure 4.9, there is perfect compensation because the time constants are identical, despite the time delay differences. Figure 4.10a demonstrates the results without compensation and Figure 4.10b shows that the compensation is limited in the presence of differing dynamics where only one coincidence point is used.

4.7.3 Extension[5]

This extension consists of delaying the initialisation of the reference trajectory by taking the delayed prediction of the disturbance into account.

The time delay of the process is r_p and r_t the delay of the disturbance $(r_p = 40 < r_t = 60)$. The disturbance pert_f is generated by a first-order, unknown process P_0, where pert_f is measured. pert_f acts through process P_2 on the output of process P_1 (*i.e.*, an additive disturbance). Thus, the CV is given by: $\text{CV} = s_p + s_{\text{pert}}$. The measured disturbance input pert_f is given by:

$$\text{pert}_f(n) = \text{pert}_f(n-1)a_f + b_f \text{Pert}(n-1),$$

where a_f and b_f simulate the unknown disturbance process P_0 .

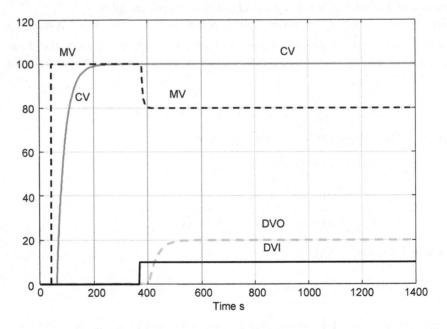

Figure 4.9. Controller with compensation: process $T_1 = 30$ s, disturbance $T_2 = 30$ s, $n_p = 20$ s, $n_{\text{pert}} = 30$

---

[5] This extension was proposed by Chris Vanbaelinghem from EVONIK.DEGUSSA.

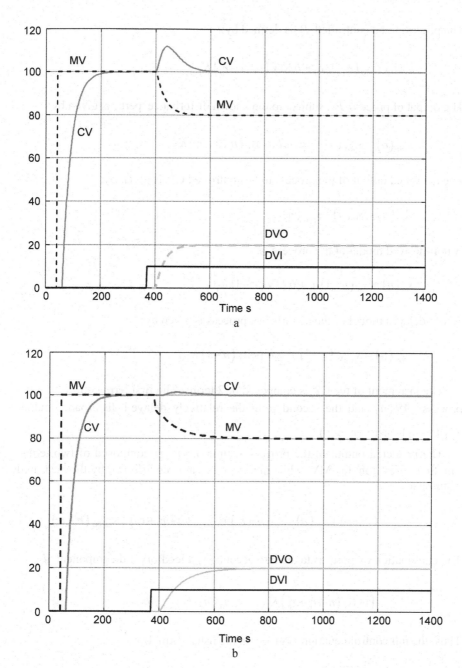

Figure 4.10. a Controller without compensation: process $T_1 = 30$ s, disturbance $T_2 = 30$ s, $n_p = 20$ s, $n_{pert} = 30$ and **b** controller with compensation: process $T_1 = 30$ s, disturbance $T_2 = 70$ s, $n_p = 20$ s, $n_{pert} = 30$

The process to be controlled P_1 is defined by:

$$s_p(\text{ii}) = s_p(n-1)a_1 + b_1\text{MV}(n-1-r_p)K_1 .$$

The output of process P_2, subject to the known disturbance pert, is given by:

$$s_{\text{pert}}(n) = s_{\text{pert}}(n-1)a_2 + b_2\text{pert}_f(n-1-r_t)K_2 .$$

The measured output of the process to be controlled CV is given by:

$$\text{CV}(n) = s_p(n) + s_{\text{pert}}(n) .$$

A non-delayed model of the process is:

$$s_m(n) = s_m(n-1)a_1 + b_1\text{MV}(n-1)K_1 .$$

A non-delayed model of the disturbance process is given by:

$$s_{\text{pert}}(n) = s_{\text{pertM}}(n-1)a_2 + b_2\text{pert}_f(n-1)K_{2M} .$$

The increment of the CV is composed of 2 terms. The first term comes from the process $\text{MV}(n)$ and the second from the relatively delayed disturbance output $s_{\text{pert}}(n-r_t)$ where $(r_p < r_t)$.

The predicted output of the process at time $n+rt$ is composed of two terms. The first arises from the MV, while the second term is contributed by the measured disturbance:

$$s_{\text{pred}} = \text{CV}(n) + s_m(n) - s_m(n-r_p) + s_{\text{pertM}}(n-r_t+r_p) - s_{\text{pertM}}(n-r_t) .$$

The contribution of pert_f is taken into account as a feedforward component d_{pert}:

$$d_{\text{pert}} = \left(\text{pert}_f(n-r_t+r_p)K_{2M} - s_{\text{pertM}}(n-r_t+r_p)\right)b_2 .$$

Thus, the full control equation is given, in the usual form, by:

$$\text{MV}(n) = \frac{(\text{Setpoint} - s_{\text{pred}})l_h + s_m(n)b_{1h} - d_{\text{pert}}}{K_1 b_{1h}} .$$

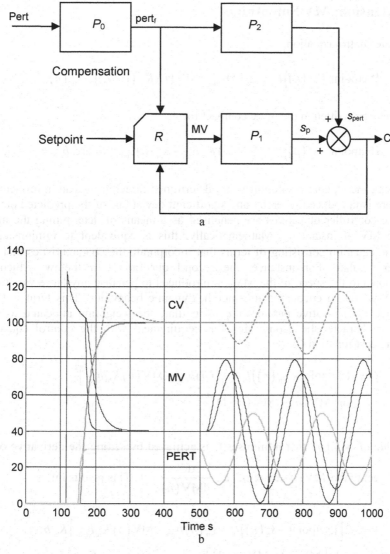

Figure 4.11. a Feedforward control with full compensation and **b** shows in solid lines control with delay compensation and feedforward compensation

The solid lines in Figure 4.11b represent control with delay compensation and the feedforward component taken into account and the dashed lines demonstrate the results without any compensation.

It is worth noting that the MV has reached its steady state value while the CV is still in dynamic transition. This behaviour is difficult to achieve using PID/Smith compensation! The procedure is slightly more complex but quite useful in certain circumstances.

4.8 Extension: MV Smoothing

Recall the control equation:

$$\left(\text{Setpoint} - s_p(n)\right)l_h = s_m(n)a_m^h + \text{MV}(n)K_m\left(1 - a_m^h\right) - s_m(n),$$

which may be written in the more compact form:

$$\left(\text{Setpoint} - s_p(n)\right)l_h = \text{MV}(n)K_m b_{mh} - s_m(n)b_{mh}, \text{ where } b_{mh} = \left(1 - a_m^h\right).$$

In general, the algebraic solution is not determined directly. Instead, a refinement procedure is introduced whereby any significant deviations of the predicted output from the controller objectives are penalised as a means of determining the most suitable MV at instant n. Mathematically, this is equivalent to minimising a quadratic equation consisting of terms that incorporate the characteristics that we wish to regulate. For instance, the second quadratic term below, which is proportional to the speed of the MV, is introduced to penalise any large variations in the MV and so encourage its smooth, dynamic behaviour. This term is then multiplied by a positive parameter q to regulate the weighting allocated to this particular characteristic. Combining this requirement with the control equation results in cost function C:

$$C = \left[\left(\text{Setpoint} - s_p(n)\right)l_h + s_m(n)b_{mh} - \text{MV}(n)K_m b_{mh}\right]^2$$
$$+ q\left[\text{MV}(n) - \text{MV}(n-1)\right]^2.$$

The value of $\text{MV}(n)$ that minimises C is achieved by setting the derivative of C with respect to the $\text{MV}(n)$ to zero, i.e., $\dfrac{dC}{d\text{MV}(n)} = 0$. This results in:

$$0 = -2\left[\left(\text{Setpoint} - s_p(n)\right)l_h + s_m(n)b_{mh} - \text{MV}(n)K_m b_{mh}\right]K_m b_{mh}$$
$$+ 2q\left[\text{MV}(n) - \text{MV}(n-1)\right].$$

Re-arranging gives:

$$\text{MV}(n) = \frac{\left[\left(\text{Setpoint} - s_p(n)\right)l_h + s_m(n)b_{mh}\right]K_m b_{mh} + q\text{MV}(n-1)}{K_m^2 b_{mh}^2 + q}.$$

Setting $q = 0$ gives the usual MV equation.

How is q selected? For $n = 0$, we have $\mathrm{MV}(0) = \dfrac{\text{Setpoint } l_h K_m b_{mh}}{K_m^2 b_{mh}^2 + q}$ compared

to $\mathrm{MV}(0) = \dfrac{l_h \text{ Setpoint}}{K_m b_{mh}}$. Smoothened $\mathrm{MV}(0) =$ unsmoothed $\mathrm{MV}(0)\beta$, where

$\beta = \dfrac{K_m^2 b_{mh}^2}{K_m^2 b_{mh}^2 + q} < 1$, this permits the computation of q as:

$$q = K_m^2 b_{mh}^2 \frac{1 - \beta}{\beta} \ .$$

Figure 4.12 shows the response of $\mathrm{MV}(n)$ for $\beta = 0.2$. Note that MV_a is "smoother", when compared to MV_{na}, as it presents a smaller initial overshoot at $n = 0$ while contributing minimal disruption to the response dynamics. The process is defined by $K = 1$, $T = 30$ s with a CLTR $= 80$ s. The measured CLTR is given by:

CLTR $= 82$ s for $\beta = 0$, and CLTR $= 83$ s for $\beta = 0.2$.

In this case, a fully calculated algebraic solution of the minimisation is achieved, without a real time iterative minimisation of the quadratic criteria. In practice, industrial practitioners generally avoid such techniques. But we will see, later in Chapter 8, that there is a good argument for the use of this procedure.

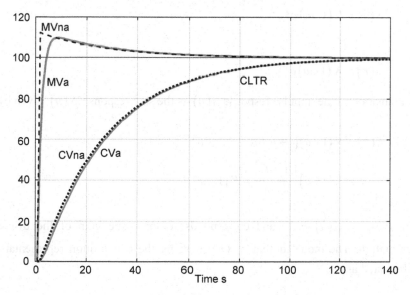

Figure 4.12. Smoothing procedure of the MV

4.9 Convolution Representation

An asymptotically stable process can be represented by a sequence of weighting coefficients a_i that may be physically interpreted as the impulse response of the process [12].

If we assume that $e(n)$ and $s(n)$ represent the process input and output, respectively, then we may describe the process using:

$$s(n) = a_1 e(n-1) + a_2 e(n-2) + \ldots + a_i e(n-i) + \ldots + a_N e(n-N) .$$

In the nominal case $s_m(n) = s_p(n)$, where $\left(a_{m_i} = a_{p_i}\right)$, N is chosen sufficiently large such that $a_N < \varepsilon$, where ε is chosen in advance. In practice, a value in the region of $N = 40$ is chosen for a system with a finite response. This form of model structure is very flexible with N degrees of freedom. The concept of order does not make sense in this context.

Remark 4.3. Consider the case of a first-order process with gain K and time constant T,

$$s(n) = s(n-1) a + K(1-a) e(n-1) .$$

If $s(0) = 0$, $e(0) = 1$ and $e(i) = 0$ for $i > 0$, we obtain:

$$s(2) = s(1) a = K(1-a) a ,$$

$$s(3) = K(1-a) a^2 ,$$

$$\ldots$$

$$s(n) = K(1-a) a^{n-1} .$$

The summation of the impulse response $s(n)$ is the step response $S(n)$ given by:

$$S(n) = K(1-a)\left(1 + a + a^2 + \ldots + a^{n-1}\right)$$

$$= K(1-a)\frac{\left(1-a^n\right)}{1-a} = K\left(1-a^n\right) .$$

When $n \to \infty$, $S(n) \to K$ and corresponds to the static gain of the discrete representation. The model output $s_m(n)$ given by the convolution representation may be stated as:

$$s_m(n) = a_1 e(n-1) + a_2 e(n-2) + \ldots + a_i e(n-i) + \ldots + a_N e(n-N) .$$

The forced and free outputs may be easily calculated. Assume that h is the coincidence point, $S_{\mathrm{L}}(n+h)$ the free output and $S_{\mathrm{F}}(n+h)$ the forced output. At instant $n+h$, the output $s_{\mathrm{m}}(n+h)$ consists of two terms, the first depends only on the future error term $e(n+i)$, where $0 \le i \le h$, and the second term depends only on the past $e(n-1)$, where $0 \le i \le N-h$.

$$
\begin{aligned}
s_{\mathrm{m}}(n+h) = {} & a_1 e(n-1+h) + a_2 e(n-2+h) + \ \ldots \\
& + a_{h-1} e\big(n-(h-1)+h\big) + a_h e(n-h+h) + \ \ldots \\
& + a_{h+1} e\big(n-(h+1)+h\big) + \ \ldots \\
& + a_N e(n-N+h),
\end{aligned}
$$

may be re-arranged as:

$$
\begin{aligned}
s(n+h) = {} & a_1 e(n+h-1) + a_2 e(n+h-2) + \ \ldots \\
& + a_{h-1} e(n+1) + a_h e(n) + a_{h+1} e(n-1) + \ \ldots \\
& + a_N e\big(n-(N-h)\big) \ .
\end{aligned}
$$

The free output may be determined by setting the future inputs to zero, $i.e.,$ $e(n+i) = 0$ for $i = 1, \ldots , h-1$ giving:

$$
S_{\mathrm{L}}(n+h) = a_h e(n) + \ \ldots \ + a_N e(n-N+h) \ .
$$

By assuming that past inputs are zero, $i.e.,$ $e(n-i) = 0$ for $i = 0 \rightarrow N-h$ and that the future inputs are assumed non-zero, the forced output may be calculated as:

$$
S_{\mathrm{F}}(n+h) = a_1 e(n+h-1) + a_2 e(n+h-2) + \ \ldots \ + a_{h-1} e\big(n+h-(h-1)\big) \ .
$$

If the future input is assumed to be a step, $i.e.,$ $e(n+i) = 1$ (unit step, elementary basis function), we obtain: $S_{\mathrm{F}}^{*}(h) = a_1 + a_2 + \ \ldots \ + a_{h-1}$.

Assuming $s_{\mathrm{m}}(n)$ is the model output; the manipulated variable $MV(n)$ is given by:

$$
MV(n) = \frac{\big[\text{Setpoint} - CV(n)\big] l_h + S_{\mathrm{L}}(n+h) - s_{\mathrm{m}}(n)}{S_{\mathrm{F}}^{*}(h)} \ .
$$

Figure 4.13 shows the open-loop response, in the form staircase of decreasing steps, and the closed-loop control response (closed loop CV_c and open loop CV_o).

The weighting sequence comprises 250 values with $a = -e^{-T_s/2}$, $h = 50$, CLTR $= 1$ (which is not significant here) and $T_s = 1$ s. At 500 s a disturbance of amplitude 10 is applied to the output. It would be difficult of regulate this process in any form other than in its convolution representation!

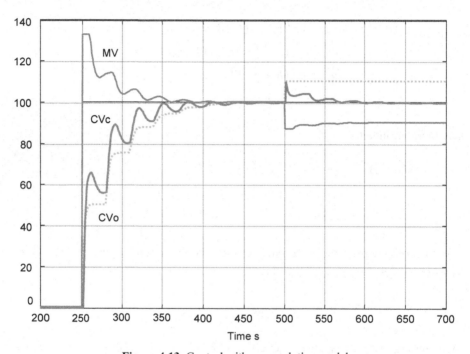

Figure 4.13. Control with a convolution model

4.10 Extension to Higher-order System Models

A process model of any order may be written in the form:

$$s_m(n) = a_1 s_m(n-1) + a_2 s_m(n-2) + \ldots + a_N s_m(n-N) + \ldots$$
$$+ b_1 e(n-1) + b_2 e(n-2) + \ldots + b_{N+1} e(n-N+1) .$$

It is assumed that the process is free of integrators. If this is not the case, the model may be modified appropriately using the decomposition principle to be described later.

4.10.1 Real Poles

If the poles of a system are first-order, real and simple, the transfer function may be decomposed into simple, first-order elements represented by $s_m(n)$, where

$$s_m(n) = s_{m1}(n) + s_{m2}(n) + \ldots + s_{mi}(n) + \ldots + s_{mN}(n),$$

and $s_{mi}(n)$ are given by:

$$s_{mi}(n) = q_i s_{mi}(n-1) + K_i(1-q_i)e(n-1) .$$

This decomposition results in the gains K_i and q_i that are related to the time constants T_i via $q_i = e^{-T_s/T_i}$.

All the modes have a *common* input variable $e(n)$ in this parallel decomposition. Thus, the control equation becomes:

$$MV(n) = \frac{(\text{Setpoint} - CV(n))l_h + S_L(n) - s_m(n)}{S_F},$$

where S_L and S_F, the free and forced outputs, respectively, are given by:

$$S_L(n) = s_{11}(n) + s_{12}(n) + \ldots + s_{1N}(n) \text{ and } S_F = s_{f1} + s_{f2} + \ldots + s_{fN},$$

with $s_{1i} = s_{mi}(n)q_i^h$, $s_{fi} = (1-q_i^h)K_i$.

It is worth noting that the coincidence point h should be chosen greater than 1 for a process of order $N > 1$ (see Chapter 6). Processes of order greater than 1 tend to possess an overshoot, as is the case with cascade control (see Chapter 7). When dealing with such systems the coincidence point should be selected sufficiently small so that the open-loop response tracks the reference trajectory.

4.10.2 Complex Poles

The computer-based controllers currently used in industry do not have complex-number-based arithmetic. Consequently, it would not be realistic to expect the procedure described in Section 4.10.1 to be used in practice. However, several other alternatives exist depending on whether the model is constant or time dependent.

One method, which is easy to implement, consists of using the decomposition principle to transform complex poles into real poles. The superposition theorem may then be applied to calculate the free and forced output responses. Recall that a process model of any order may be represented by the finite-difference equation:

$$s_m(n) = a_1 s_m(n-1) + \ldots + a_N s_m(n-N) + \ldots + b_1 e(n-1) + \ldots + b_M e(n-M)$$

where $(M \le N - 1)$.

4.10.2.1 Calculation of the Free Solution S_L

If the controller is to be adaptive, varying in time with the parameters of the model, the free mode must be explicitly computed in real time. Consider the example of a second-order process (see Figure 4.14) defined by:

$$s_m(n) = a_1 s_m(n-1) + a_2 s_m(n-2) + ke(n-1) .$$

At n, a homogeneous equation of the form $s_{ml}(p)$ exists such that :

$$s_{ml}(p) = a_1 s_{ml}(p-1) + a_2 s_{ml}(p-2) + 0 .$$

For a coincidence point h, $s_{ml}(p)$ is iterated for $p = 3, 4, \ldots, h-1$, with initial conditions $s_{ml}(2) = s_m(n-1)$, $s_{ml}(1) = s_m(n-2)$. The free solution $S_L(n)$ is then obtained using $S_L(n) = s_{ml}(h-1)$.

Note that this procedure may be applied to all non-stationary (time-varying) processes defined by a stable finite-difference equation.

4.10.2.2 Calculation of the Forced Solution S_F

Assume we have a stable, open-loop, integrator-free[6] process. The procedure consists of assuming that the regulator is in steady state with a non-zero set-point producing a solution $[MV_0, CV_0]$, where $CV_0 = KMV_0$ and K is the steady state model gain. Given the Setpoint $= CV_0$, the control equation may be stated as:

$$MV_0 = \frac{\left(0 + MV_0 \sum S_{me} + K.MV_0 \sum S_{ms}\right)}{S_F}$$

where the summations $\sum(S_{me})$ and $\sum(S_{ms})$ consist of the free, past outputs of the finite difference model $s_m(n-i)$ and the past input values of the $MV(n-i)$, where $S_F = K - \sum S_{me} - K \sum S_{ms}$. Consider the case of a first-order system where:

---

[6] If an integrator is present in the process, the decomposition principle can be applied.

$$MV(n) = \frac{\left(\text{Setpoint} - CV(n)\right)l_{\text{h}} + s_{\text{m}}(n)\left(1 - a_{\text{m}}^{h}\right)}{K_{\text{m}}\left(1 - a_{\text{m}}^{h}\right)},$$

or alternatively: $MV(n) = \dfrac{\left(\text{Setpoint} - CV(n)\right)l_{\text{h}} + s_{\text{m}}(n)\left(1 - a_{\text{m}}^{h}\right)}{S_{\text{F}}}.$

In steady state, $s_{\text{m}0} = K_{\text{m}}MV_0$ and $MV_0 = \dfrac{0 + s_{\text{m}0}\left(1 - a_{\text{m}}^{h}\right)}{S_{\text{F}}}$, from which we obtain

$$s_{\text{m}0} = \frac{K_{\text{m}}s_{\text{m}0}\left(1 - a_{\text{m}}^{h}\right)}{S_{\text{F}}} \text{ and } S_{\text{F}} = K_{\text{m}}\left(1 - a_{\text{m}}^{h}\right). \text{ Substituting for } S_{\text{F}} \text{ gives:}$$

$$MV(n) = \frac{\left(\text{Setpoint} - CV(n)\right)l_{\text{h}} + s_{\text{m}}(n)\left(1 - a_{\text{m}}^{h}\right)}{K_{\text{m}}\left(1 - a_{\text{m}}^{h}\right)}.$$

Remark 4.4. In order to prevent steady state errors, resulting from numerical inaccuracies, it is preferable to calculate the forced output in a manner consistent with the calculation of the free output, rather than integrating the difference equation subjected to a unit step input.

Example 4.1. Consider the case of a second-order system given by:

$$s_{\text{m}}(n) = a_1 s_{\text{m}}(n-1) + a_2 s_{\text{m}}(n-2) + ke(n-1),$$

where $a_1 = 1.9844$, $a_2 = -0.9850$, $k = 6.2683/10,000$. Note that no oscillatory mode is present in the closed-loop response when $h = 35$ and CLTR $= 400$ s.

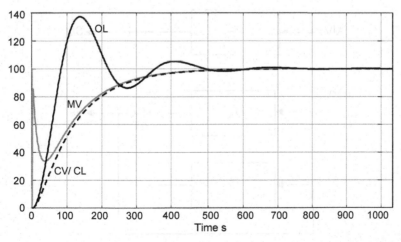

Figure 4.14. Control of a second-order process with complex poles, *i.e.*, open loop (OL)/closed loop (CL)

4.10.3 Control of a Pure Oscillator

Industries such as defence, aeronautics and, to a lesser extent, robotics, that deal with advanced technology on a daily basis, have evolved a more "holistic" philosophy when designing new products from the beginning. Such an approach is referred to as *integrated design*. This methodology takes the control specifications into account from the outset of the project and results in systems that may be controlled more effectively, giving better performance. Typical systems designed in this manner have a different physical structure and exhibit unstable or lightly damped oscillatory behaviour that must be stabilised artificially. In such cases, the systems are not viable in open loop, and the application of control techniques is essential.

Is it possible to control a pure oscillatory process without using derivatives?

The second-order transfer function process $H(s) = \dfrac{k\omega_n^2}{s^2 + 2\zeta\omega_n s + \omega_n^2}$ represents an oscillator when ζ is practically zero.

This problem may be treated in a manner similar to the method shown in Section 4.9. However, it is preferable to decompose the model using the decomposition principle as it may be applied to all processes; albeit asymptotically stable, marginally stable or unstable.

The decomposition principle is used as shown in Figure 2.4 where the forward path is taken as a second-order process with time constants T_1 and T_2. The feedback path also consists of a second-order system but with a different gain. An example is presented in Figure 4.15 as it demonstrates the basic generalisation procedure.

The process is given by $H_p(z^{-1}) = \dfrac{kz^{-1}}{1 - q_1 z^{-1} + \sigma q_2 z^{-2}} = \dfrac{CV(z^{-1})}{MV(z^{-1})}$.

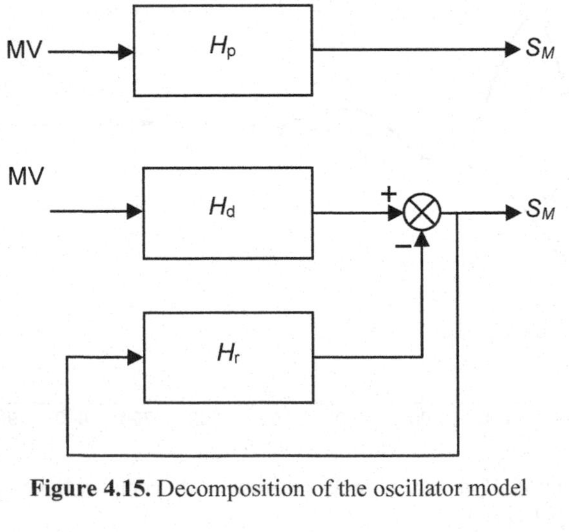

Figure 4.15. Decomposition of the oscillator model

The forward path is given by:

$$H_d\left(z^{-1}\right) = \frac{kz^{-1}}{1 - q_1 z^{-1} + q_2 z^{-2}} = \frac{S_1\left(z^{-1}\right)}{MV\left(z^{-1}\right)}$$

and the feedback path by:

$$H_r\left(z^{-1}\right) = \frac{-Gz^{-2}}{1 - q_1 z^{-1} + q_2 z^{-2}} = \frac{S_2\left(z^{-1}\right)}{CV\left(z^{-1}\right)}$$

where $a_1 = e^{-\frac{T_s}{40}}$, $a_2 = e^{-\frac{T_s}{60}}$, $k = 0.75\left(1 - q_1 + \sigma q_2\right)$.

A process, such as $H_p\left(z^{-1}\right)$, is chosen where $q_1 = a_1 + a_2$, $q_2 = a_1 a_2$, $\sigma = 1.005$ and $T_s = 1\,\text{s}$. For $\sigma = 1.0427$ the process represents a system of diverging oscillations. A value of $\sigma = 1.0426$ generates a characteristic response as shown in Figure 4.16. $\left(MV = 100\right)$. The model transfer function is of the form:

$$S_M\left(z^{-1}\right) = MV\left(z^{-1}\right)H_d\left(z^{-1}\right) + S_M\left(z^{-1}\right)H_r\left(z^{-1}\right).$$

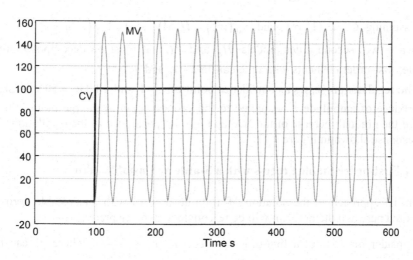

Figure 4.16. Open-loop oscillator

Substituting for $H_d\left(z^{-1}\right)$ and $H_r\left(z^{-1}\right)$ and solving for $S_M\left(z^{-1}\right)$ we get:

$$S_M\left(z^{-1}\right) = \frac{kz^{-1}}{1 - q_1 z^{-1} + q_2 z^{-2} - Gz^{-2}} MV\left(z^{-1}\right).$$

Assuming $G = q_2(1-\sigma)$, and $S_M(z^{-1}) = H_p MV(z^{-1}) = CV(z^{-1})$ we find that there are two, second-order models H_d and H_r that must be decomposed into simple, first-order elements with time constants T_1 and T_2.

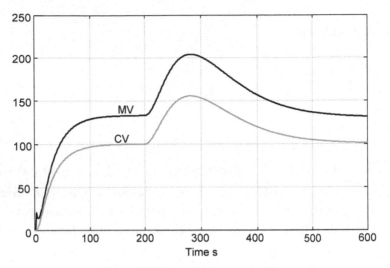

Figure 4.17. Control of a pure oscillator

Assuming $G = q_2(1-\sigma)$, and $S_M(z^{-1}) = H_p MV(z^{-1}) = CV(z^{-1})$ we find that there are two, second-order models H_d and H_r that must be decomposed into simple, first-order elements with time constants T_1 and T_2.

The regulator possesses a coincidence point of $h = 1$ and CLTR $= 50$ s. A step disturbance of amplitude 3 is applied to the process input at sample point 200 s. Notice the gradual increase of the MV and the disappearance of the oscillations in the response shown in Figure 4.17.

4.10.4 First-order Stable Process with a Stable or Unstable Zero

An initial approach to the solution of a stable process with an unstable zero is presented here and the problem will be re-considered in Chapter 9.

Consider the case of a first-order process $H_1(s) = \dfrac{1 + sT_1}{1 + sT_2}$, where T_1 can be positive or negative and T_2 is always positive. In the case shown in Figure 4.18 we assume $T_1 = -50$ s and $T_2 = 100$ s. In order to facilitate simulation, $H(s)$ is taken as:

$$H_1(s) = A + \frac{B}{1 + sT_2}, \text{ where } A = \frac{T_1}{T_2} \text{ and } B = 1 - A .$$

The output may be decomposed into two parallel processes S_{1m} and S_{2m}. Thus, the control equation will include these two terms:

$$S_{2m}(n) = S_{2m}(n-1)a_2 + (1 - a_2)B.MV(n-1) .$$

The input MV must be delayed by one sampling period (zero-order hold). The free output of S_{1m} is zero, whereas its forced output is given by: $S_F = A.MV(n)$. The control equation is given by:

$$(\text{Setpoint} - CV(n))l_h = S_{2m}(n)a_2^2 + B(1 - a_2^2)MV(n)$$
$$- S_{2m}(n) + A.MV(n) - S_{1m}(n-1),$$

from which $MV(n)$ is extracted:

$$MV(n) = \frac{(\text{Setpoint} - CV(n))l_h + S_{1m}(n) + S_{2m}(n)b_{2h}}{A + Bb_{2h}}, \text{ where } b_{2h} = 1 - a_2^2 .$$

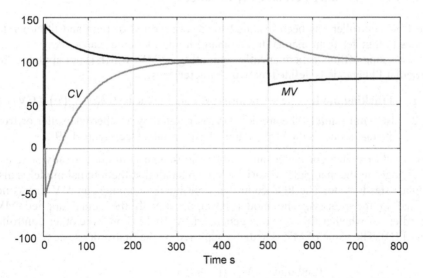

Figure 4.18. Control of a zero-order process $H = (1 - s50)/(1 + s100)$ with a stable pole and an unstable zero

It is advisable to choose a coincidence point h such that the value of $CV(n)$ for a step input possesses the same sign as that of the process gain. In order to determine the critical coincidence value h_c, an open-loop test may be carried out with

$MV(n) = 1$. The value for h_c should be chosen such that the regulated output $CV(n)$ evolves beyond the "dip" in the inverse response (where $h_c = 43$).

Note that the regulated output $CV(n)$ possesses an inverse response and that the system loop is satisfactory (see Figure 4.18) despite an additive disturbance on the CV at 500 s. In this case, $TRBF = 120$ s and $h = 150$ are chosen.

4.11 Controller Initialisation

It is a relatively simple exercise to transfer from an automatic to manual mode of regulation. But, switching from a manual to automatic mode is always a little agonising for operators, particularly when the system contains an integrator.

Generally, the objective is to provide a smooth transition between the MVs at the point of switching. This eliminates the possibility of a discontinuity arising in the manipulated variable, *i.e.*, a "bumpless" procedure. Two situations exist where a bumpless transfer is required: (a) the PFC controller is permanently installed and (b) the PFC controller is in a commissioning phase.

4.11.1 PFC Controller Permanently Installed

The PFC controller has been installed for a long period of time and the operator wishes to transfer from manual to automatic mode for some reason. Any installed PFC, working offline, is permanently computing its MV, which is not applied. This particular PFC mode is defined by two characteristics:

1. **Tracking mode:** The set-point is set equal to the CV, *i.e.*, $S_p(n) = CV(n)$.
2. **Internal model:** The applied MV, generated by another controller or from manual mode, is the input to the internal model (see Figure 4.19).

It is usual to switch controller mode when the system is in quasi-steady state and the changes in the manipulated variable are so small that the internal model is also in quasi-steady state. The PFC controller continually computes an MV that is not applied to the process. The model is supplied with the actual applied MV; regardless of whether this signal is generated by the PFC or some other controller (*e.g.*, PID). The value of MV is actually supplied by the first-order system:

$$MV_{pfc}(n) = \frac{(\text{Setpoint} - CV(n))l_h + s_m(n)b_m}{K_m b_m} \ .$$

Since the Setpoint $= CV(n)$ and $s_m(n) = K_m MV(n)$, we get:

$$MV_{pfc}(n) = \frac{s_m(n)}{K_m} = MV(n) \ .$$

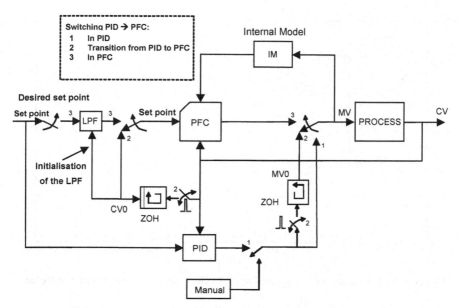

Figure 4.19. Switching from PID or manual modes to PFC control

Thus, at the time of switching from manual control to PFC control there is no "bump" as the $MV_{pfc}(n)$ is equal to the $MV(n)$. However, it is probable that the local set-point will not match the desired set-point. Consequently, the subsequent values of $CV(n)$ should converge on the desired set-point. This correction may be carried out, without any special considerations, if the $(Setpoint - CV)$ difference is small.

However, if the $(Setpoint - CV)$ difference is large (see Figure 4.20), the desired set-point convergence may be achieved using the output of a first-order filter. First, the filter output is initialised to the existing process output value $CV(n)$. Then, the input of the filter is switched to the desired set-point value and the filter output is used as the transition set-point profile. At the time of switching Setpoint $= CV(n)$ results in a bumpless transition from the current set-point to the desired true set-point.

The only precaution to be observed is that the system should be in steady state for a time period longer than the model OLTR to ensure that the applied MV is in quasi-steady state. The switching operation may be performed by the operator without any particular difficulty.

Note that a PFC controller never stops computing the MV. This is the case regardless of whether the computed MV is applied to the process or not. But, the internal model of the PFC controller is supplied by the MV that is *actually* applied to the process, whatever its origin (PFC controller or not).

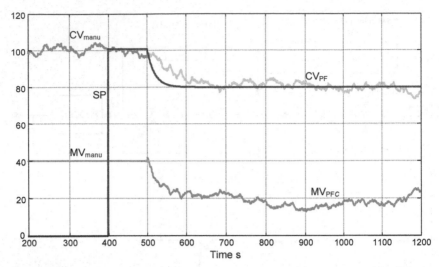

Figure 4.20. Initialisation of the predictive controller with a large set-point/CV difference

4.11.2 PFC Commissioning Phase

The strategy described above may still be applied when a PFC controller is installed for the first time in the control system. However, to reach the steady state more rapidly, we can momentarily divide the model dynamics (time constant, for instance) by a factor of 10 or 100. In this situation, the model will react more rapidly than the original model, and will appear quasi steady state. It is sufficient to change this dynamic during one sampling period only. This procedure may be easily implemented since the dynamic of the model (*i.e.*, the time constant) is not a fixed parameter but a variable. Thus, a programmed variation of the time constant of the internal model may be introduced.

A final remark: this bumpless switching procedure is an absolute requirement if operators are to accept this control technique!

4.12 Summary

- The future $MV(n+i)$ where $(1 < i < H)$ are projected onto a set of polynomial basis functions whose coefficients are computed to generate future $CV(n+i)$ that coincide with the reference trajectory.
- A solver is used to find the parameters of the polynomials. The selection of the solver is open: a simple algebraic solution is sufficient if the order of the polynomial is less than 3.
- No tracking error exists when following a polynomial set-point of order N if the functional basis is of order $N - M$, where M is the number of integrators of the process.

- PFC can be implemented in practice on different processes: integrative, delayed, unstable, higher order.
- Switching back and forth from any controller (manual) to PFC is bumpless.

5

Tuning

Abstract. In this chapter we address the tuning procedure. The objective is to obtain the desired degree of accuracy, dynamics and robustness required by a process in order to fulfil its operational specifications. These three requirements represent the fundamental objectives of regulation. The question is how the 3 PFC tuning parameters: basis function, reference trajectory and coincidence horizon are able to cope with these 3 requirements. The issue of trade-offs between the different specifications is discussed. The concepts of gain time, delay margin and sensitivity function are described. The chapter concludes by presenting a practical tuning aid for first-order systems.

Keywords: Accuracy, dynamics, robustness, gain margin, time-delay margin, sensitivity function, tuner's rule

5.1 Regulator Objectives

The specifications given to all regulators may be characterised by these three requirements [9]:

Accuracy: The set-point, either static or dynamic, should be followed without error. This should continue to be the case regardless of any perturbations or regulator mismatch that might arise subsequent to the initial commissioning of the regulator. However, this requirement can prove more difficult to maintain in the presence of a dynamic set-point, *e.g.*, a ramp set-point is more difficult to track than a constant set-point.

Dynamics: In the time domain, where this requirement is generally defined, dynamics may be described as a fixed-time response to a change in set-point or a disturbance. However, it may also be characterised in the frequency domain in terms of a fixed-frequency disturbance rejection.

Robustness: Robustness quantifies the deterioration in performance with respect to the stated regulation objectives arising from process structural variation. This requirement is satisfied at the time of the regulator's design.

The ideal regulator should possess the ability to satisfy the accuracy, dynamics and robustness requirements independently, by adjusting a specific parameter of the controller. Such a regulator is referred to as a "diagonal" regulator, which is not the case in PID. In the real world, this ideal does not exist. In fact, as the following examples will show, the emphasis may be put either on dynamic performance (*e.g.*, fast servomechanisms of constant structure) or on stability (*e.g.*, distillation columns for various types of crude oil). Difficulties arise when these requirements are in conflict, thus it is necessary to identify a compromise between them.

5.2 Accuracy

Accuracy is dictated by the choice of basis functions (see Section 4.3). This requirement does not pose any major problems provided that the process is stable in closed loop.

For example if a process with no integrator has a parabolic set-point, a basis polynomial $F(n) = u_0 + u_1 n + u_2 n^2$, where u_0, u_1 and u_2 are unknown, may be used.

Various set-point trajectories may be generated using the form $\text{Cons}(n) = K n^J$. For example, $J = 0$ produces a step set-point and $J = 1, 2$ and 3 generate ramp, quadratic and cubic set-points, respectively. Set-points higher than third order are rarely used in practice.

The PFC controller uses a combination of the above trajectories to generate a generalised function basis set, *i.e.,*

$$F(n) = \sum_0^J u_k n^k \text{ , for } k = 0, 1, \dots, J \text{ .}$$

Note that if the process contains an integrator it is necessary to reduce the order of the basis polynomial by the number of integrators present. For example, if the process contains one integrator in the presence of a quadratic set-point, the basis function consists of two components (a step and a ramp), since a step and ramp result in a quadratic output $F(n) = u_0 + u_1 n$.

Note that it is necessary to choose at least one basis function, $F(n) = u_0$, with the capacity to generate a non-zero MV at time 0. It is mathematically possible to satisfy a long-term objective, *e.g.*, $h = 10$, by determining a combination of the remaining nine MVs without acting at time 0. Since this calculation is repeated at each sampling point the resulting MV values applied at each sampling point have the same problem, and the resulting MV will remain permanently at zero!

The accuracy is fixed by only one parameter − the order of the basis functions. Therefore, accuracy is achieved without compromising the other specifications; this is a unique attribute of predictive control.

5.3 Dynamics

The dynamics of the closed-loop system are dictated by the closed-loop controller. The controller should be able to follow a change of set-point with a specified closed-loop time response and be able to reject, in the regulation mode, the disturbances acting on the process. Tracking-mode tuning is carried out in the time domain, while in the process mode tuning is carried out in the frequency domain using the transfer-function representation.

5.3.1 Time Response

In Chapter 2 we saw that the choice of the CLTR determines the dynamic closed-loop response. The CLTR must be chosen in accordance with the stated system specification. However, the system response must be compatible with the process' natural, open-loop response and, in particular, its OLTR. If the CLTR < OLTR at the time of set-point change, the MV will overshoot, *i.e.*, the intermediate dynamic values will exceed the final value and, in turn, will run the risk of compromising the physical speed and amplitude constraints of the system. On the other hand if the CLTR > OLTR, the MV will generally remain constrained so that the response will remain lower than its final value (see Figure 5.1).

Consider the example of a unity gain, third-order process with time constants $T_1 = 10$ s, $T_2 = 40$ s, $T_3 = 60$ s and a coincidence point of $h = 15$. Figure 5.1 shows the system response as the CLTR progresses from 10 s to 80 s. Note that for a CLTR = 45 s, the output is practically identical to that of the open-loop case.

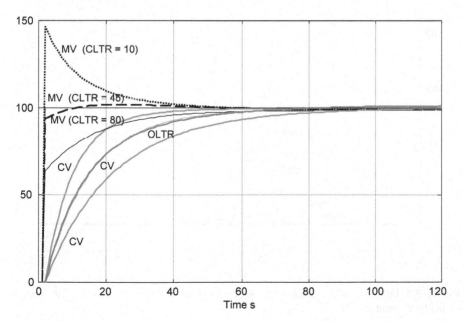

Figure 5.1. Time response of MV/CV in open loop and in closed loop for different CLTRs

The CLTR is not, in itself, significant. In fact, it is the CLTR/OLTR ratio that is important. Generally, in the case of steady state, open-loop processes, the smaller the CLTR/OLTR ratio, the weaker the robustness of the process becomes. We begin to see the effects of the tuning (*i.e.*, dynamics – robustness compromise) referred to in Section 5.1. It is here that the designer's skill is called upon to select the most appropriate compromise as a solution.

It is worth noting that fixing the value of the time constant of the closed loop *a priori*, by specifying the reference trajectory, does not result in a fixed CLTR. In fact, even in the *nominal* case, where the model and process are assumed identical, it is necessary that a coincidence point of $h = 1$ is chosen to achieve a stable regulator.

The reference trajectory must be re-initialised at each sample point on the actual measured value of the process output. It is possible to implement such a procedure for a first-order process. However, this is not generally the case for processes of a higher order because h must be chosen greater than 1. If the desired and resulting CLTR are similar and, more importantly, change in the same manner, this does not pose any practical problems when tuning.

We will use a third-order process with a OLTR $= 12$ s to help us understand the influence that the coincidence point h has on the CLTR. The desired closed-loop time response is CLTRD and the real closed-loop time response is given by CLTRR .

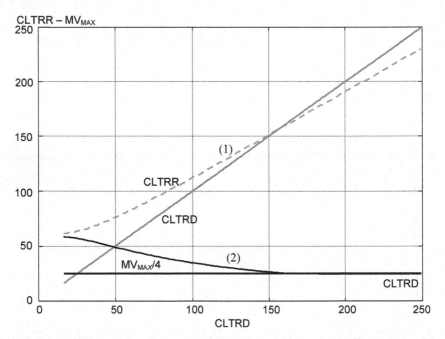

Figure 5.2. Influence of the desired CLTR (CLTRD*)* on the real CLTR (CLTRR) with a coincidence point $h = 30$

Figure 5.2 $(h = 30)$ and Figure 5.3 $(h = 10)$ show the effect that the choice of coincidence point for a CLTRD has on the resulting CLTRR and the maximum value of MV for the desired CLTR (CLTRD).

The smaller the value of the coincidence point h becomes, the closer the value of the effective closed-loop time response CLTR is to the required CLTRD. However, the initial MV will increase in magnitude.

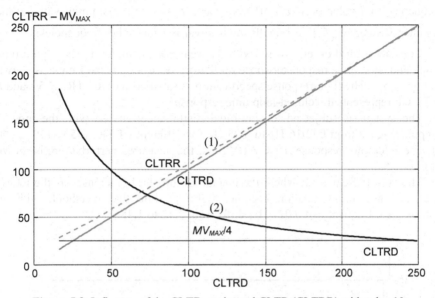

Figure 5.3. Influence of the CLTR on the real CLTR (CLTRR) with a $h = 10$

5.3.2 Frequency Response

In practice, the frequency response is studied to analyse the degree of (unmeasured) disturbance rejection. This may be represented in simulation by adding a sinusoidal signal DV at a particular frequency to the process output. In fact, the disturbance may be inserted at any point in the process but an equivalent disturbance signal injected at the output may always be found (see Figure 5.4).

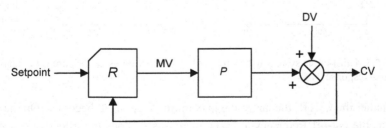

Figure 5.4. Equivalent frequency disturbance model

If the disturbance has very low frequency, effectively approximating a constant signal, it will be completely rejected since there is no positional error with PFC. As is the case with most physical systems, the open-loop response of the process, *i.e.*, $OL = R.P$ (see Figure 5.4) is low-pass. Consequently, as the frequency increases toward infinity, all finite MVs will produce no output and so the disturbance will not be attenuated. The value $A = \dfrac{\text{amplitude of CV}}{\text{amplitude of DV}}$ increases, with increasing frequency, to a value of 1 (cut-off frequency, F_c) up to a maximum value at the resonant frequency (F_r) and finally tends towards 1 for higher frequencies.

Consider a third-order system (with one real pole, complex poles, a pure time delay and an inverse response) whose oscillatory open-loop response OL is shown in Figure 5.5. The time-response specification is satisfied with CLTR = 75 s and h = 25. CL represents the closed-loop time response.

The system is subjected to sinusoidal disturbances mixed with the process output, ranging from 0.0016 Hz to 0.05 Hz for different CLTRs (75 s, 150 s, 300 s). The Bode plot response $A = \text{CV/DV}$, *i.e.*, the measured perturbation, is shown in Figure 5.6.

The cut-off frequency, where the magnitude of $A = 1$, increases as the CLTR decreases and the regulation becomes tighter, while the overshoot, with an approximate frequency of 0.021 Hz, shifts from 1.12 to 1.45.

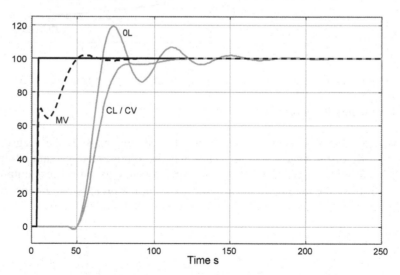

Figure 5.5. Third-order system with a real pole, complex poles and a pure time delay with an inverse response

The smaller the CLTR, the larger the maximum value of A becomes. On the other hand, as the cut-off frequency $(A = 1)$ increases so does the effectiveness of the rejection at low frequencies (Figure 5.6).

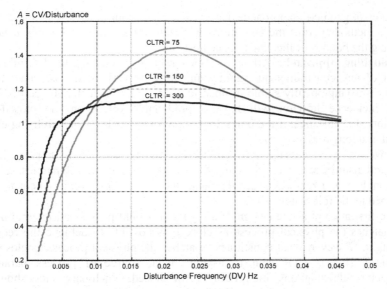

Figure 5.6. Harmonic disturbance rejection

The characteristic in Figure 5.6, with its distinctive "hill" shape, is referred to as the *sensitivity function*. The sensitivity function indicates the limited ability of a system to reject disturbances over its entire frequency range of operation. For instance, it is not possible to have good disturbance rejection at low frequencies without introducing an amplification of the medium-frequency range[7]. Therefore, the choice of closed-loop time response must be made while taking the inevitable presence of disturbances into consideration. If the frequency spectrum of the expected disturbances approaches the summit of the "hill", then the use of this regulator would be ill-advised. In fact, the system response would be better if no regulation were used at all!

Thus, the use of the closed-loop time response to correct the overall system time response has implications for the system's ability to reject disturbances.

5.4 Robustness

The most important property of a regulator is its stability. Above all, a regulator should not cause any harm and should ensure secure operation when all operating variables lie within acceptable limits. The stability of the nominal regulator, *i.e.*, where the model and process are identical, represents a problem that can be approached in two ways:

Traditional approach: In the case of linear unconstrained processes the regulator, as described in Chapter 3, is linear. Consequently, stability may be determined using classical methods (*e.g.*, pole placement, algebraic criteria,

---

[7] This is sometimes referred to as the "waterbed" effect.

frequency or geometric approaches) as used in linear control theory. In some cases, specific methods permit the conditions sufficient for stability to be met, *a priori*, by fixing the bounds of the coincidence horizon.

Modelling approach: All applications require modelling to determine the impact of any constraints on stability and may be verified *a posteriori*. If the process is asymptotically stable in open loop, it is clear that the choice of CLTR and coincidence horizon will have a predictable effect on the closed-loop behaviour. For instance, if the poles are real, a long CLTR has a stabilising effect, as well as a long horizon.

However, as necessary as stability is, on its own it is not sufficient. Structural disturbance, *i.e.*, change of process gain, delay, dynamics, *etc.* is commonplace and requires a detailed analysis. The stability requirement may be described in the following terms:

The first attempt at the design of a regulator should possess an internal model that matches the physical process as closely as possible to achieve the required regulation. However, what constitutes "matches the physical process as closely as possible"? This is equivalent to asking "what degree of process parameter mismatch results in an unstable process"? And indeed, which parameters should be considered in making such a determination?

Any parameter may be chosen. But, in general, it is the parameters that require the least adjustment, which results in unstable behaviour, that are of most interest. Traditionally, two approaches are considered when it is necessary to determine the degree of parameter variation that results in instability:

Gain margin (GM): This is the multiplicative gain factor that renders the process unstable. Generally, for normal systems a large gain is usually a contributory factor in causing instability. In practice, a GM greater than or equal to 2 is recommended.

Delay margin (DM): This represents the positive or negative variation in process delay that renders the system unstable. There are two DMs for each CLTR. It is clearly in the practitioner's interest to increase the DM (*e.g.*, transmission delay of data, delays due to actuators) in accordance with the tuning parameters.

Note that is possible for the phase margin to increase, and in certain situations this causes the delay margin to decrease. In fact, the use of phase margin, historically taught in classical courses in control, should be avoided. The above definitions have the benefit of being easily understood by all.

Any attempt at regulation should start with an investigation of the process-stability margin requirements. These requirements may vary substantially: Some processes have constant structures and, in such cases, a high dynamic performance may be expected.

In other processes, the structure is highly variable and only modest dynamic performance can be expected. If the performance is not acceptable, it is advisable to adapt the internal model to the known structural process variations *a priori*. It is worth noting that most of the predictive controllers installed in industry possess *variable internal models*: chemical reactors with variable load, mechanical servos with variable load, *etc.* Some industrial examples will be described in Chapter 10.

If we re-examine the third-order process used previously, the following relationship between the CLTR and the multiplicative gain factor of the process

GM, which renders the closed loop unstable, may be traced for a given coincidence horizon (see Figures 5.7 and 5.8). The OLTR is 65 s and is indicated by the "*" in both diagrams.

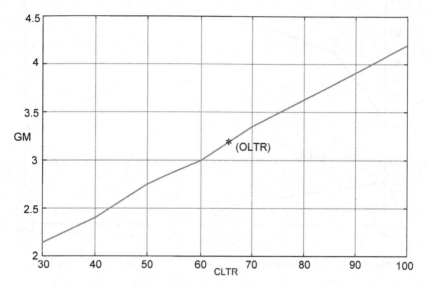

Figure 5.7. GM – CLTR relationship

A practical approach consists of starting with an evaluation of the possible domain of potential structural variations of the process to determine the stability margins for a given regulator. If the robustness is compatible with the structural variations, a stationary regulator structure is retained; otherwise, it is necessary either to increase the CLTR or to search for an adaptive physical model with measurable, physical parameters capable of altering the structure.

5.5 Choice of Tuning Parameters

5.5.1 Accuracy

The choice of basis functions is straightforward. In the PFC case, the basis functions are represented in polynomial form but other choices are possible. The maximum polynomial degree is selected based on the nature of the set-point and the degree of persistent disturbances.

As already stated, it is always necessary to have a step component as one of the basis functions. Expanding the basis set to include polynomials of a higher degree than necessary has little practical benefit.

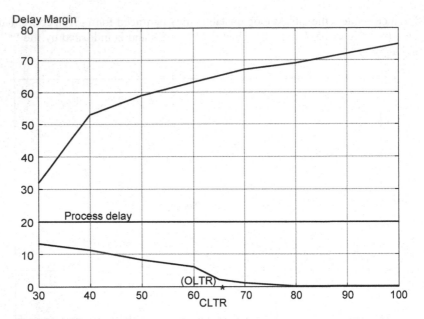

Figure 5.8. Positive and negative delay margins as a function of the CLTR

5.5.2 Dynamics

The closed-loop system dynamics are determined by setting the CLTR. The value of the CLTR, in turn, dictates the reference trajectory and influences the cut-off frequency and the resonant frequency. It is worth bearing in mind that, depending on the choice of coincidence horizon, the resulting CLTR is generally longer than required, assuming the horizon is long. Also, the choice of CLTR assumes that the coincidence horizon is compatible with the desired dynamics (see Section 5.4.3).

5.5.3 Robustness

Robustness depends not only on the dynamics but also on the choice of coincidence point. This statement has several implications, as we have already seen in Section 5.4.

5.5.3.1 Manipulated Variable Activity
Consider the third-order system discussed previously (see Section 5.3.1) and observe the effect that the choice of h has on the coincidence horizon.

Short h: As noted previously in the nominal case, the output at the current sample point corresponds exactly with that of the reference trajectory. If we assume that there are no constraints on the process, the output at the next sampling point also correlates with the reference trajectory. As the same conditions exist at subsequent sampling points, the process outputs track the initial reference trajectory and the process loop behaves like a first-order system. The cost of such precise control is that the resulting manipulated variable MV is very dynamic. In contrast, the behaviour of the CV is smooth because it is close to the exponential reference trajectory.

Long h: When $h \leq$ OLTR, the forced output response for a step basis function is the static gain of the process that is assumed asymptotically stable.

In this case the MV closely resembles a step with little variation and no oscillation. In contrast, the resulting CV is similar to the open-loop response, which may be very dynamic. Depending on the circumstances, these two cases present clear benefits, depending on the control specifications.

Figure 5.9 summarises the effects of h on the behaviour of the MV and CV at a change in set-point.

h	MV	CV
Short	Dynamic	Smooth
Long	Step type	Open-loop behaviour

Figure 5.9. Effect of h on MV and CV

5.5.3.2 Turpin Point

At a change in set-point the first value of the MV is a good indicator of the "nervousness" of the regulator. So, from an operational perspective, a pertinent question that may be posed at this point is: For a given CLTR, is it possible to choose a coincidence point that produces a $\mathrm{MV}(0) = \mathrm{MV}_0$ minimum? At instant 0, it is assumed that the system is at state 0 with all variables initialised to zero. Thus, the control equation reduces to:

$$\mathrm{MV}_0 = \frac{(\mathrm{Setpoint} - 0)I_h}{\mathrm{Forced\ Output}} \ .$$

In the normalised case of a unit set-point $(\mathrm{Setpoint} = 1)$ becomes:

$$\mathrm{MV}_0(h) = \frac{1.\left(1 - e^{-\frac{3hT_s}{\mathrm{CLTR}}}\right)}{\mathrm{Forced\ Output}(h)} \ .$$

In the third-order process case, for the variable h, the function $\mathrm{MV}_0(h)$ passes through a minimum (see Figure 5.10). This point is referred to as the *Turpin point*[8] and is situated before the peak of the characteristic response that maximizes the denominator but does not minimise MV_0. This position varies somewhat with the CLTR, which is a general property of this parameter.

---

[8]   P. Turpin was the technical manager of Compagnie Générale d'Electricité.

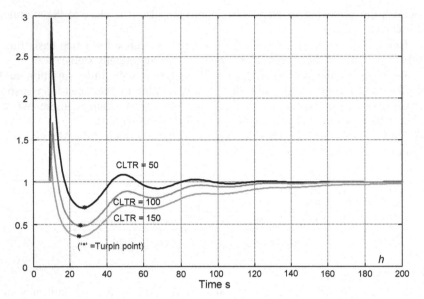

Figure 5.10. Turpin point

5.5.3.3 Inverse Response (Non-minimum Phase)

In practice, it is not uncommon to encounter processes that, when subjected to a step input, produce a response $s(t)$ of the opposite sign to that of the final response. Such a system is termed an inverse response or a non-minimum phase process. This situation arises in cases such as the control of the height of water in a steam drum, in some distillation-column tray temperatures and in flexible mechanical systems.

Assume we have an asymptotically stable system with a positive gain in open loop. When the process is subjected to a positive set-point the response follows the profile given by:

$$0 < t < t_0, \quad s(t) < 0,$$

$$t = t_0, \quad s(t_0) = 0$$

and $t > t_0, \quad s(t) > 0$.

It is advisable, under these conditions, to ensure that the choice of coincidence point is located beyond t_0, i.e., $hT_s > t_0$. While the CV will finally achieve its objective, it (surprisingly) proceeds in the "wrong" direction at first before correcting to the expected direction. If the controller is not aware of this surprising behaviour this could result in the system response becoming unstable. In Section

5.5.3.2, during our study of the Turpin point, we saw that h may be varied beyond the value of h_0 $(i.e., h > h_0)$.

Recall that in the case of a system with *pure time delay* the choice of h is always made without taking the delay into account. The delay is incorporated by shifting the initialisation of the reference trajectory's origin by an amount equal to the magnitude of the delay (see Section 3.3 and Appendix A.1).

5.5.3.4 First-order System Case

The impulse response of a stable system $\text{Simp}(n)$, of order $N > 1$, is always zero. Thus, for $n = 0$, $\text{Simp}(0) = 0$. For a first-order system $N = 1$, with time constant T, CV is a maximum at $n = 1$ and decreases exponentially with a decrement $a = e^{-\frac{T_s}{T}}$. Recall from the process control equation that $e(n)$ represents the manipulated variable and $s_p(n)$ represents the regulated output of the process:

$$s_p(n) = s_p(n-1)a + K(1-a)e(n-1),$$

where the model is assumed identical to the process, *i.e.*, $s_p(n) = s_m(n)$. Eliminating $e(n)$ from the process equation $s_p(n)$ by converting $e(n)$ to $e(n-1)$ in the control equation and substituting into $s_p(n)$ gives:

$$e(n) = \frac{\left(\text{Setpoint}(n) - s_p(n)\right)l_h + s_m(n)\left(1 - a^h\right)}{K\left(1 - a^h\right)},$$

where $l_h = 1 - e^{-\frac{3hT_s}{\text{TRBF}}} = 1 - \lambda^h$. In the nominal case, we get,

$$s_p(n) = s_p(n-1)d + (1-d)\text{Setpoint}(n-1),$$

i.e., the closed-loop relationship between the set-point $\text{Setpoint}(n)$ and the output $s_p(n)$. The output $s_p(n)$ has unity gain and dynamics $d = 1 - (1-a)\frac{l_h}{b_h}$ that may be re-written as $d = 1 - (1-a)\frac{1 - \lambda^h}{1 - a^h}$. The value of d depends on $\lambda(\text{CLTR})$ and h, resulting in a degree of freedom. So, it is possible to find a combination of the required CLTR and h to satisfy the equation for $s_p(n)$.

If, for simplicity, we choose $h = 1$ and $d = 1 - (1 - \lambda) = \lambda$ and if the model and process are identical, we get a first-order transfer function between the set-point and the CV of unity gain and dynamics λ with effective CLTR = required CLTR.

5.6 Gain Margin as a Function of CLTR (First-order System)

Assume that the process and model dynamics are identical and that their gains K_p and K_m are different. How does this gain mismatch affect the robustness? Using the process equation from Section 5.5.3.3 and defining $R = \dfrac{K_p}{K_m}$ we get:

$$s_p(n) = s_p(n-1)q + (1-q)\text{Setpoint}(n),$$

where $q = \dfrac{1 - R(1-a)(1-a^h)}{l_h}$. There is no positional error but the response dynamics depend on the CLTR, h and R. Figure 5.11 shows the variation of the effective CLTR (CLTR_R) with respect to the process gain K_p for $h = 1$, $K_m = 1$ and the required $\text{CLTR} = 100\,\text{s}$. Then, $K_m = 1$, $\text{CLTR\_R} = \text{CLTR\_SPEC}$.

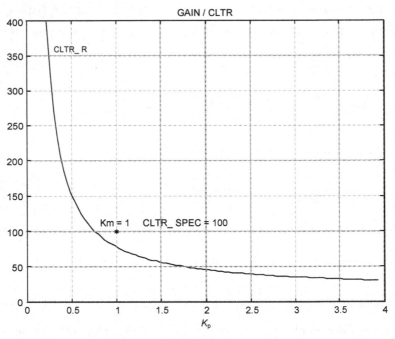

Figure 5.11. CLTR_R $/ (K_p / K_m)$

5.7 Tuning

Assume, for simplicity, that we have only one basis function – a step. The only tuning parameters are:

- the desired CLTR;
- the position of the coincidence point h (only one).

There are several criteria to fulfil:

- **Steady state accuracy:** No permanent offset error will result, even in the presence of permanent state disturbance (additive perturbation) or model mismatch (internal model $\neq$ process), given that the closed-loop system is asymptotically stable and an independent model is used.
- **Dynamic response:** depends mainly on the selected CLTR but the selection of the coincidence point has some influence for models of order greater than 1. For a step change in set-point the first manipulated variable value $MV(1)$ produced may have a value much higher than its final value and this phenomenon should be taken into consideration.
- **Dynamic perturbation rejection:** This must be analysed in the frequency domain, *e.g.*, at what frequency does resonance occur and what is its corresponding amplitude?
- **Trade-off:** The ideal controller would be a controller whose tuning parameters tune only one of the above specifications. In the case of a PID controller, the 3 parameters (P-I-D) have an influence on all the aforementioned characteristics.
- **CLTR/h:** The selection of CLTR and h is carried out by the following mathematically analysis. For a small h, coincidence point, the actual CLTR is close to the desired CLTR. But the first manipulated variable value $MV(1)$ becomes quite large if the CLTR is less than OLTR. We will demonstrate this using a third-order process with real poles (T_1, T_2, T_3) and a zero T_n.

The input data are $T_1 = 20, T_2 = 40, T_3 = 60, T_n = 10$ with a sampling period $T_s = 1$ (normalised value). The open-loop time response OLTR = 245 and inflexion point HH = 36.

The values selected from Figure 5.12 are $H = 75$, $CLTR_{desired} = 180$. The result is:

- GM = 69.7/delay margin = 118;
- Max. sensitivity function = 1.12;
- Period = 131.13 and $CLTR_{actual} = 209$.

For a set-point = 100: $MV_{max} = 130.7$ (with a process gain =1).

Referring to Figure 5.13 the open-loop step response has an "s" shape with an inflexion point at time HH. The value of h should be selected in the vicinity of HH.

Three values of h around HH and several desired CLTR values around the open-loop time response OLTR are tested. The selection of h and CLTR involves a trade-off between the magnitude of the resulting $MV(1)$ and the desired CLTR, *e.g.*, if a small value for h is selected this implies that $MV(1)$ will be large (see Figure 5.12).

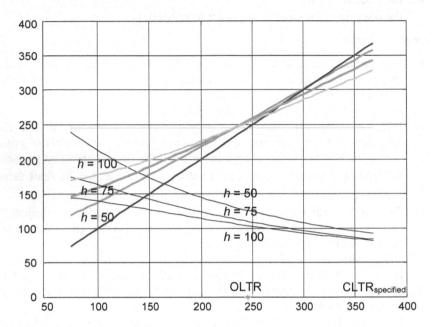

Figure 5.12. $CLTR_{actual}$ and first $MV/CLTR_{specified}$ for $h = 50,\ 75$ and 100, *e.g.*, $h = 75$, $CLTR_{specified} = 180$, $CLTR_{actual} = 209$ and $MV(1) = 130.7$

5.7.1 Gain Margin/Delay Margin

The gain and delay margins are computed in the frequency domain. The transfer function is decomposed in 3 elementary first-order processes. The time scale is normalised by setting the sampling period $T_s = 1$. There is no time delay in the process.

The computed time-delay margin will be only taken as an increase of delay. However, in some cases, instability may result if the process time delay is smaller than the model time delay!

We consider that the set-point is zero and the analysis is carried out using discrete transfer functions, *i.e.*, z-transform. The classical PFC control equation in the z-domain is given by:

$$MV(z) = \left((0 - CV(z))l_h + MV(z) \sum H_i(z) b_i \right) \Big/ \sum K_i b_i \ .$$

(see Section 4.10.2.) When the GM is required:

$$CV(z) = MV(z)H_p(z)GM,$$

and when the time-delay margin is sought: $CV(z) = MV(z)H_p(z)e^{-j\omega DM}$. At the frequency of instability: $\sum K_i b_i = H_p(z)GMl_h + \sum H_i(z)b_i$. Defining $Q(z) = \sum K_i b_i - \sum H_i(z)b_i$ we get:

$$GM = Q(z)/Hp(z)GMl_h = R(z) .$$

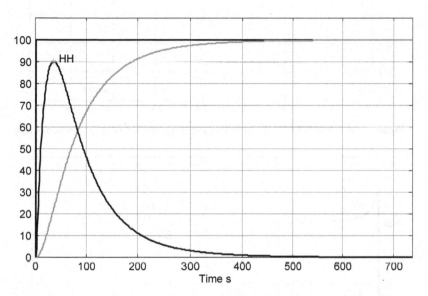

Figure 5.13. Open-loop step response. Inflexion point: HH = 36, OLTR = 245

For a frequency ω_0: $\text{imag}(R(z)) = 0$ and $GM = \text{real}(R(z))$.

Similarly, for a frequency ω_1: $(z_1 = j\omega_1)$, $\cos(\omega_1 DM) = \text{real}(R(z_1))$ and $\sin(\omega_1 DM) = -\text{imag}(R(z_1))$. The solution for z_1 is such that $\left[\text{real}(R(z_1))\right]^2 + \left[\text{imag}(R(z_1))\right]^2 = 1$ and DM is given by:

$$DM = \frac{1}{\omega_1}\text{acos}(R(z_1)) .$$

5.7.2 Sensitivity Function ("Hill Curve")

Feedback control is flawed! In most cases frequencies exist where, if disturbances occur, the control system increases the disturbance effect. In such cases it would be better not to apply any control at all!

Figure 5.14. Sensitivity function ("hill curve"): mod(CV/DV)/Disturbance period

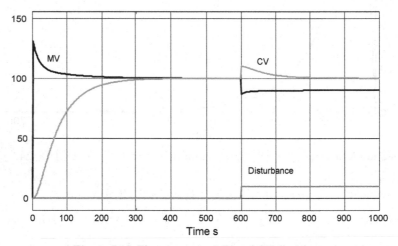

Figure 5.15. Time response: MV and CV disturbance

As discussed in Section 5.4, the CLTR has a strong impact on the sensitivity function. In the case of the CLTR = 180 s and $h = 75$ s we obtain the results shown in Figure 5.14. Note that the shorter the CLTR, the higher the peak of the sensitivity function. But, for low frequencies the disturbance rejection is better. So, it is clear that knowledge of the disturbance spectrum should be exploited in the selection of the CLTR. Figure 5.15. demonstrates the time-domain response of the system to a step disturbance with a step disturbance taken into account.

5.8 The Tuner's Rule

It is common to encounter the situation where an operator sits at a monitoring station in the control room with only qualitative views of the MV and CV responses of say, a first-order process with a gain K_p and time constant τ_p available. There is no access to on-line system identification methods or data-logging facilities. However, the operator does possess the ability to alter the gain and time constants of the internal model of the first-order process. Under these circumstances the operator may alter the model parameters, using observations of the MV and CV curves, until satisfactory results are achieved.

If the settling time of the process is relatively short, *i.e.*, of the order of minutes, the following procedure may be used to tune the system with a high cost/benefit ratio. However, for longer settling times, it is better to use the classical off-line system identification approach. The practical tuning suggestions here assume the following conditions have been met:

- A model of the process has been estimated.
- The controller has been installed.
- The system has been validated in steady state.
- The regulator has made its first stable dynamic control test.

At this point, we proceed with caution and choose a CLTR = OLTR. This minimises the risk of instability due to internal model mismatch. Note that we are trying to identify the OLTR of the unknown process with an objective of modifying the OLTR of the model (time constant).

- **Target:** We consider a first-order process defined by a gain K_p, time constant τ_p and time delay D_p to be determined experimentally by visual inspection of the process response. An open-loop test gives an initial visual approximation of the values of these parameters: K_{m0}, τ_{m0} and D_{m0}.

A set of PFC tests with internal models K_m, τ_m, and D_m are performed with CLTR = OLTR = $3\tau_{m0}$. Note that the OLTR is evaluated without the time delay and represents the time taken for the process output to reach 95% of its final value. The complete "settling time" is then the sum of the time delay plus the OLTR.

- **Identification of the gain** K_m: We fix $\tau_{m0} = \tau_0$, $D_{m0} = D_0$ and we have assumed that $K_p > 0$. A positive step set-point of PFC is applied with a specified CLTR = OLTR, the first $MV(1)$ is smaller or bigger than the final steady state $MV(\infty)$ for a given gain K_m to be identified:

  1. If $MV(1) > MV(\infty)$: the gain K_m should be increased
  2. If $MV(1) < MV(\infty)$: the gain K_m should be decreased

3. If $K_m = K_p$ the initial and final MVs are equal: this value for K_m is selected as $K_m^* = K_m$. Note that this is valid even if $\tau_{m0} \neq \tau_0$ and $D_{m0} \neq D_0$. (see Figures 5.16a–c.)

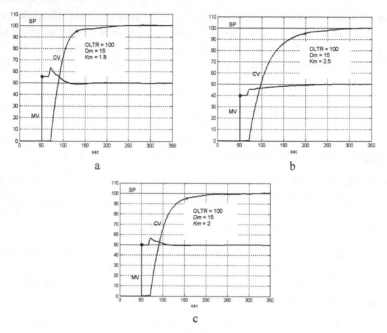

Figure 5.16. Identification of K_m where **a** $MV(1) > MV(\infty)$, **b** $MV(1) < MV(\infty)$ and **c** $MV(1) = MV(\infty)$

*   **Identification of the time delay (CLTR = OLTR):** $K_{m0} = K_m^*$, $\tau_m = \tau_{m0}$, CLTR $= 3\tau_{m0}$ are fixed during the search of D_m and we have assumed that $K_p > 0$. Assume that a step is applied at time t_0. At time $t_0 + D_m$, the $MV(n)$ increases linearly until time $t_0 + D_p$. After this time the MV starts controlling the process smoothly.

1. If the slope of $MV(1)$ is positive: D_m should be increased.
2. If the slope of $MV(1)$ is negative: D_m should be decreased.
3. If $D_m^* = D_p$ the $MV(n)$ starts controlling at time: $t_0 + D_m^*$ Note that this is valid even if $\tau_m \neq \tau_{m0}$ but with an already identified gain K_m^*. (see Figures 5.17a–c.)

- **Identification of the time constant (CLTR varies as OLTR varies):** $K_m = K_m^*$, $D_m = D_m^*$ are kept constant during the identification of τ_m. But, the CLTR varies with the different OLTR or τ_m under test, where OLTR = CLTR = $3\tau_m$. Generally, the $MV(n)$ looks like a step.

  1. If $MV(n)$ demonstrates a rise, OLTR is increased.
  2. If $MV(n)$ demonstrates a dip, OLTR is decreased.
  3. If $MV(n)$ is constant $\tau_m : \tau_m$ = OLTR/3 (see Figures 5.18a–c).

Generalisation. If the CV is above (below) the set-point, increase (decrease) the parameter. This statement applies to the 3 parameters.

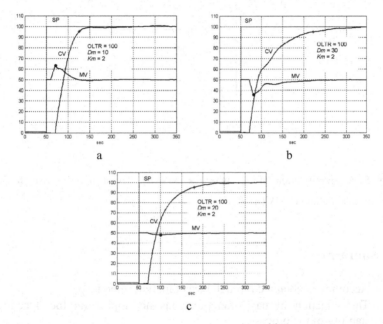

a

b

c

Figure 5.17. Identification of D_m where **a** the slope of $MV(1)$ is positive, **b** the slope of $MV(1)$ is negative and **c** $D_m^* = D_p$

5.9 Practical Guidelines

- **Accuracy:** Fixed by the choice of the basis functions and independent of the other parameters.
- **Dynamics:** Fixed by the CLTR. Depends on the coincidence horizon if h is long.
- **Robustness:** Depends on the CLTR and choice of the coincidence horizon.

In the common case of a first-order process with pure time delay and step set-point there is only one parameter of regulation, *i.e.*, the CLTR, whose physical significance is intuitive and obvious.

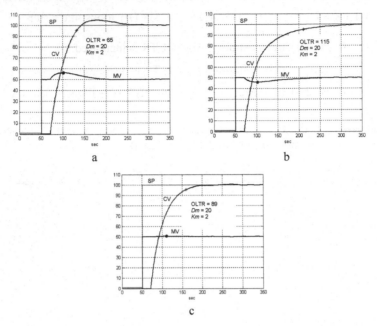

a

b

c

Figure 5.18. Identification of τ_m where **a** $MV(n)$ demonstrates a rise, **b** $MV(n)$ demonstrates a dip and **c** $MV(n)$ is constant

5.10 Summary

- Accuracy is obtained through basis function selection.
- The selection of basis functions has no impact on the dynamics and stability of the process.
- Dynamics in the time domain is defined by the reference trajectory time response.
- The sensitivity function in the frequency domain may be easily computed.
- Robustness is obtained for a given set of dynamics by choosing appropriate coincidence point(s).
- If the coincidence horizon is short, the MV is quite active but the predicted response is close to the reference trajectory.
- If the coincidence horizon is long, the MV is less active but the CV differs from the reference trajectory.
- The gain and time delay margins are readily accessible.
- For first-order processes a tuner's rule, based on simple observations of the CV behaviour, is proposed.

6

Constraints

Abstract. Handling constraints on MVs and internal variables of the process is an important issue when implementing a PFC controller. In this chapter an approximate, sub-optimal solution is proposed, which is sufficient in most cases and easy to implement.

Keywords: MV constraints, CV constraints, CLTR constraints, multiple controller

6.1 Benefit

The necessity to optimise production, in terms of quantity or quality, requires complete control over all the action variables of the process. Such optimisations can result in some process variables approaching their physical limits. These limitations or constraints may arise from:

- **Actuators:** Amplitude, speed, power limitations while running a servomechanism or the flow limitations of a heat exchanger.
- **Internal variables of the process:** A situation that is regularly encountered in heat-treatment systems involves controlling the temperature of a process while respecting a separate temperature constraint on another process that also depends on the same temperature controls.

The situation may also arise that the constraint on the MV does not have to be applied rigorously but may be voluntarily relaxed in order to protect the process from excessive actions.

Respecting constraints has practical benefits. All regulators have some form of physical constraints imposed on their actuators that, if violated, would reduce the life span of the system components. Also, accounting for constraints on state variables of a process can lead to large economic savings.

If a constraint exists, a controller should not produce a manipulated variable that would produce a response limited by the physical constraints of the system, which may or may not be accurately known. A typical example is the clipping of the output control voltage caused by amplifier saturation.

It is imperative to operate in the linear region and implement decisions that respect all current and future constraints. The effectively applied action should be *explicitly constrained a priori* by the controller. The internal model should be fed by the constrained input, which will not violate the process' domain of validity, resulting in an accurate prediction. The resulting performance would almost certainly be less efficient than having no constraints, but the regulation would remain both stable and effective.

This approach results in a linear system, while the overall controller becomes non-linear. From a theoretical perspective, this represents a difficult problem [4, 11, 13, 18]. This dilemma may be easily appreciated when one considers that, in order to satisfy a particular objective, it is necessary to test all future scenarios at every sample point in the future. Such an approach is necessary because any decision taken at the current sample point must not only respect the constraints at that sample point but must also prevent situations that result in the process violating any future constraints. Unfortunately, it is not possible to implement this "omnipotent" approach to optimal control theory. It would be too difficult to implement such a solution in industrial automation applications using current control technology.

Is it possible to find a local solution that represents a satisfactory compromise between ease of implementation, while delivering an acceptable sub-optimality? The simplest approach to taking amplitude constraints on the MV into account is to pass the projections of the MV produced by the regulator through appropriate limiters. It is these constrained $\mathrm{MVL}(n)$ values and not the calculated $\mathrm{MV}(n)$ that are supplied to the internal model of the regulator whose output is SML (see Figure 6.1). When considering constraints on the dynamic internal process variables X_i, it is necessary to ensure that the projection of the $\mathrm{MV}(n+j)$ at the current sample point is compatible with the constraints imposed on the variables X_i. It would be necessary to consider all future scenarios $(n+j)$ at all future instants n that are impossible in practice. Thus, we must be content with considering the future scenario for the current sample point only and to repeat this strategy at each future sampling point, and that turns out to be quite sufficient for most industrial applications.

6.2 MV Constraints

In general, only the most common level constraints MV_{max} and MV_{min} and speed constraints D_{max} and D_{min} are taken into account (see Figure 6.1).

$$\mathrm{MV}_{min} \leq \mathrm{MV}(n) \leq \mathrm{MV}_{max}, \text{ and } D_{min} \leq \left[\mathrm{MV}(n) - \mathrm{MV}(n-1)\right] \leq D_{max} .$$

The MVC calculated by the regulator is initially passed through a speed limiter followed by an amplitude limiter. The resulting value $\mathrm{MVL}(n)$ is then supplied as the input to the internal model of the regulator that, in turn, produces the model output SML.

Figures 6.2 and 6.3 show the effects of including and neglecting constraints on the behaviour of the regulated first-order process with $T = 30$ s, $K = 1$, Setpoint = 100 and a CLTR = 45 s subject to the following constraints:

$$MV_{max} = 120, \quad MV_{min} = -120, \quad D_{max} = 2, \quad D_{min} = -2 .$$

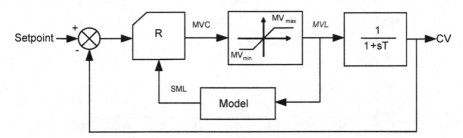

Figure 6.1. The inclusion of level and speed constraints

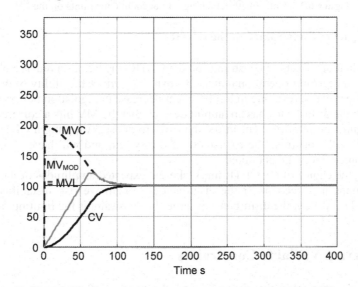

Figure 6.2. Control taking into account constraints on the MV

It can be clearly seen that the behaviour caused by the inclusion of the constraints (Figure 6.2) differs considerably from that of the unconstrained case (Figure 6.3). In both cases the manipulated variable is subject to a physical constraint. This constraint may be taken into consideration in the calculation of the control. For instance, in Figure 6.2, the model input is $MV_{MOD} = MVL$; while in Figure 6.3 $MV_{MOD} = MV$, where MV is the manipulated variable calculated by the PFC block. The failure to account for the constraint *a priori* results in an unacceptable overshoot of the regulated variable.

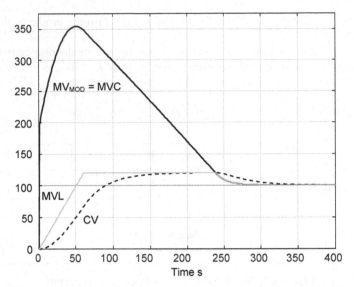

Figure 6.3. Control without taking into account constraints on the MV

6.2.1 Impact of Constraints on the CLTR

The debilitating effects of constraints on the CLTR may be observed as a change in set-point or regulator operational mode. Constraints reduce the ability of the MV to control the process, resulting in a subsequent decrease in regulation performance.

When the disturbance has an amplitude such that the MV hits a constraint there is no significant improvement in the dynamic response. Adopting a CLTR that is too small will jeopardise the robustness of the system and will have no effect on the dynamics since MV is limited.

Thus, the choice of the CLTR must take the expected amplitudes of the process perturbations into account. If the disturbance is small we risk instability with a short CLTR, while if the disturbance is large the constraint is the limiting factor.

6.3 Internal Variable Constraints

Assume that our objective is to control the process P_1, using a manipulated variable MV that also acts on the process P_2, subject to the constraint $CV_2 < CONT_2$.

This strategy, referred to as the "multiple controller" technique, consists of two regulators operating in parallel with the choice of MV being made by some form of logical, decision-making process, *e.g.*, an if-then rule set, fuzzy logic, artificial neural network.

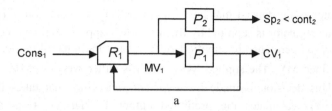

a

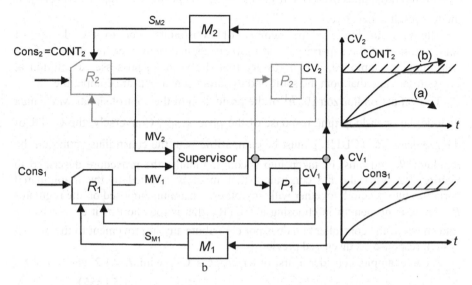

b

Figure 6.4. a Control without the need for constraint and **b** control with a constraint on a variable

A classical regulator R_1 calculates a future projection of the manipulated variable to satisfy the set-point $Cons_1$ while respecting all the constraints in this loop that act on the manipulated variable MV_1 (max, min, speed, *etc.*). The effect that the $MV_1(n+i)$ has on the future output of the constrained process CV_2 during a particular horizon $[0, h]$ is calculated. Two possible outcomes may result:

- The predicted output CV_2 respects the constraint and, in this case, the $MV_1(n+i)$ projection is acceptable and so $MV_1(n)$ is applied (see Figure 6.4 a).
- The predicted output violates the constraint and MV_1 cannot be accepted (see Figure 6.4 b).

So, what value of manipulated variable MV_2 would change CV_2 while respecting the constraint $CONT_2$? Again, we are faced with a problem of regulation.

A virtual regulator is introduced that has $\mathrm{Cons}_2=\mathrm{CONT}_2$ as a set-point. The internal model of this regulator is supplied by the actual MV applied to the process P_2. The regulator R_2 is permanently supplied by CV_2 and the internal model is supplied by the retained MV. The applied MV, which is either MV_1 or MV_2, is decided *a priori*. This decision is made by a *logical supervisor* that takes the projection $MV_1(n+i)$, calculates the predicted output $CV_2(n+i)$, tests the constraint and applies $MV_1(n)$ if the test result is positive or applies $MV_2(n)$ if the test result is negative.

There is no limit to the number of constraints, nor to the degree of sophistication of the supervisor, that may be applied. However, regardless of the configuration employed, it is necessary that the problem possesses a solution at every instant and that both the static and dynamic constraints are compatible.

The validation horizon $[0,\ h]$ of the projection of the control signal MV_1 must include the open-loop time response of the process P_2. Conversely, the CLTR of the regulator R_2 $\left(\mathrm{CLTR}_2\right)$ must be compatible with the constraints acting on the regulator R_1 and must be taken into account. Finally, the procedure described in Section 6.2 will take the constraints that must be applied to the manipulated variables into account, together with any other constraints imposed on the regulator R_1. There is no benefit in choosing a CLTR_2 that is too short as it decreases the robustness of the controller and does not contribute any improvement to the overall system response, as discussed previously.

As an example, consider a first-order process CV_1 with $K=1.2$ and $T=40$ s and a second-order process CV_2 with overshoot $H(s)=\dfrac{(1.5+s55)}{(1+s30)(1+s20)}$. In this case, it is probable that the output would exceed the constraint if the MV constraints procedure of Section 6.2 were not implemented. These processes are controlled by two regulators with $\mathrm{CLTR}_1=70$ s and $\mathrm{CLTR}_2=25$ s.

Referring to Figure 6.5, at 134 s the MV controlling CV_2 is switched from MV_2on to MV_1on. The switch drives CV_2, which is currently at saturation, towards the desired set-point value of 100. Also, note that a speed constraint exists, *i.e.*, $dMV(n)=4$, and that it has an effect on both MV_1 and MV_2. The passage from MV_2 to MV_1 is smooth as the mode change occurs only when both intersect.

Although sub-optimal, this procedure is frequently applied as it is easily implemented while offering potential economic benefit. In the specific case where the processes are of a similar nature, an override procedure [10] may be used to produce the same results; although, in the override procedure each controller is not aware of the other's behaviour. Note that the whole future bahaviour of the process P_2 is effectively computed and a real prediction is made.

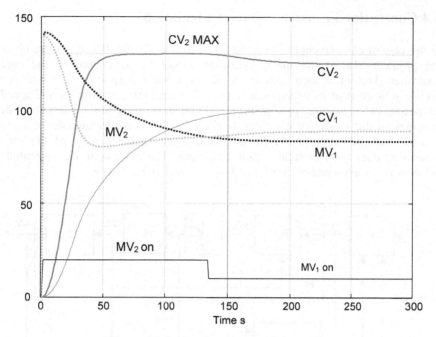

Figure 6.5. Respecting the constraint on CV_2

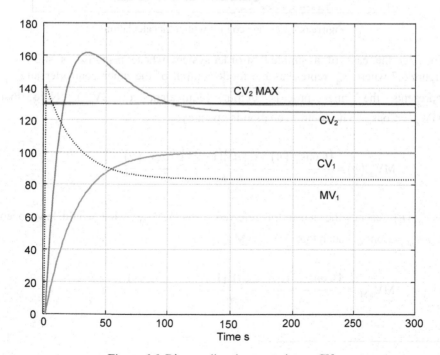

Figure 6.6. Disregarding the constraint on CV_2

6.4 Constraint Transfer – Back Calculation[9]

In the case of cascade control (see Section 7.3) it is crucial that the contraints on the physical MV of the inner controller should be transferred to the outer controller. The inner loop contains a physical actuator (pump, *etc.*) that may routinely function at its operational limits (*e.g.*, max./min. amplitudes and speed) restricting the regulation performance and the outer regulator's internal model becomes invalid. Consequently, "transferring" this constraint to the outer regulator ensures that the set-point presented to the inner regulator (MV of the outer controller) does not violate any such conditions. The inner controller essentially behaves as though it had been informed of the restriction *a priori*.

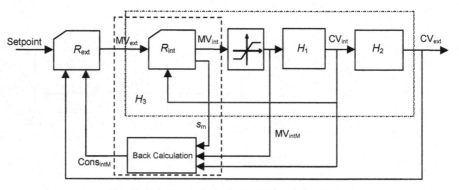

Figure 6.7. Cascade control with back calculation

Consider the case of a simple first-order system whose response is shown in Figure 6.7 where s_m represents the model output of the inner controller and s_p represents the inner process output (equivalent to CV_{int}). Note that $MV_{ext} = Cons_{int}$. The control equation is given by:

$$MV_{int}(n) = \frac{\left(Cons_{int}(n) - s_p(n)\right)\left(1 - \lambda^h\right) + b_m s_m(n)}{K_m b_m}.$$

$MV_{int}(n)$ reaches the constraint, referred to as MV_{intM}: what would have been $Cons_{int} = Cons_{intM}$ such that $MV_{int} = MV_{intM}$?

$$MV_{intM} = \frac{\left(Cons_{intM}(n) - s_p(n)\right)\left(1 - \lambda^h\right) + b_m s_m(n)}{K_m b_m},$$

---

[9] The most appropriate term "back calculation" was proposed by Heiko Luft of EVONIK.DEGUSSA.

where the variables $s_p(n)$ and $s_m(n)$ are available at each sample point. Thus, we get:

$$\text{Cons}_{\text{intM}} = s_p(n) + \frac{\text{MV}_{\text{intM}} K_m b_m - b_m s_m(n)}{1 - \lambda^h} .$$

This value supplies the outer controller's internal model:

$$s_m(n) = s_m(n-1) a_m + K_m b_m \text{Cons}_{\text{intM}}(n-1) .$$

Thus, the objective is achieved, *i.e.*, the constraint has been respected and the internal model is supplied with a "good" value at the next sample point $\left(\text{Cons}_{\text{int}}(n-1)\right)$.

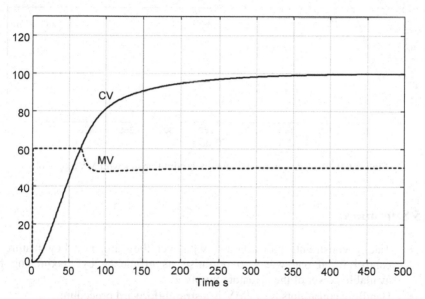

Figure 6.8. Cascade control with constraint transfer: $\text{MV}_{\text{MAX}} = 60$

In certain cases, not transferring the constraint may result in catastrophic behaviour. The constraint $\text{Cons}_{\text{intM}}$ is variable because of $s_p(n)$ and $s_m(n)$ and so must be calculated at each sample point n. Also, it may be added to the other constraints calculated by the external regulator. Figure 6.8 demonstrates the effect of a transfer-function constraint in the nominal case where the internal model of the first-order system is identical to the process.

- Process H_1 $K_1 = 1$, $T_1 = 20$ s .
- Process H_2 $K_2 = 2$, $T_2 = 60$ s .

The internal model of the external regulator R_{ext} is given by $K = K_2 \times 1$, $T_3 = 80$ s and the time constants of the regulators are $\text{CLTR}_1 = 15$ s and $\text{CLTR}_2 = 120$ s. The set-point is adjusted to 100 and the manipulated variable is constrained at 60. Note that a significant overshoot is observed when the transfer function constraint is omitted (see Figure 6.9).

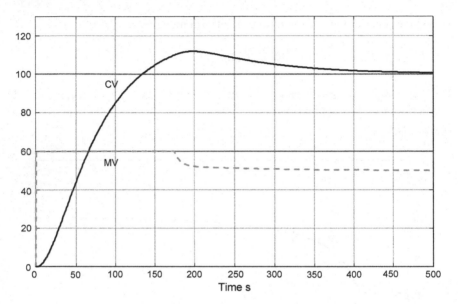

Figure 6.9. Cascade control without constraint transfer with $\text{MV}_{\text{MAX}} = 60$

6.5 Summary

- Taking constraints into account, whatever they are, is a key feature of control when one wants to optimise its efficiency by maximising the available power of the actuators involved.
- Handling constraints on an MV is a straightforward procedure.
- Constraints on an internal variable are approached by the multiple-controller technique that makes a scenario test between prospective future MVs and ultimately selects the MV that respects the internal constraint.
- If a constraint appears in the inner loop of a cascaded control, this constraint is transferred to the outer controller – a procedure that is sometimes referred to as "back calculation". This approach ensures that the internal model of the outer controller is fed by the correct MV, which is an important issue in practice.

7

Industrial Implementation

Abstract. In industrial control, it is common to utilise a configuration of several controllers to solve a particular control problem. The configuration adopted is usually application-specific. This chapter deals with the configurations most commonly encountered in practice: feedforward, transparent control, cascade control, split range, zone control, control of one and two CVs with two MVs, *etc.*

Keywords: Zone control, cascade control, constraint transfer, back calculation, transparent control, shared multi-MV control, coupled actuator, split-range control, estimator, feedforward, non-linear, scenario method, 2MV/2CV

7.1 Implementation

There are many industrial examples where predictive control has been implemented with one manipulated variable MV, one regulated variable CV and several feedforward variables. However, in the majority of cases, predictive control is implemented in conjunction with other controllers such as PID or auxiliary predictive regulators [16, 17].

The use of an internal model of the process, regardless of its complexity and with the capacity to dissociate the free and forced outputs, offers great flexibility during the implementation phase. This approach enables controller designs to be tailored to meet any application-specific requirements that may arise.

We will now consider the most common controller configurations that are encountered in practice. This discussion is intended to be representative of the possible configurations and not to be considered as an exhaustive survey. Such configurations include:

- zone control;
- cascade control;
- transparent control;
- shared multi-MV control;
- coupled actuators;

- split-range control;
- estimator;
- non-linear control;
- scenario control;
- 2 MV/2CV control.

7.2 Zone Control

The use of zone control is ubiquitous in industry, because it is simple to implement and easy to tune. The zone-control approach involves varying the CLTR as a function of the regulation error $\varepsilon(n) = \text{Setpoint} - \text{CV}(n)$.

It is often the case in industrial process control applications that the desired set-point is not a single, definitive value. In fact, the set-point is more likely to consist of a band (or zone) of values in which the CV is to be maintained, *i.e.*, $\text{Setpoint} + \text{DCVM} \geq \text{CV}(n) \geq \text{Setpoint} - \text{DCVm}$.

If the absolute error value is large, *i.e.* in the region of $\text{Setpoint} + \text{DCVM}$ or $\text{Setpoint} - \text{DCVm}$, it is advisable to adjust the CV in the direction of the set-point quickly. Hence, a small CLTR is desirable under such circumstances. On the other hand, if the error is within the prescribed zone, a large CLTR (but not infinite) is desirable. If the signs of the disturbances are unpredictable *a priori,* remaining in the centre of the zone reduces to a minimum the probability of touching the zone limits. The CV must return, albeit slowly, toward the set-point.

This approach possesses many potential benefits:

- If the frequency spectrum of the disturbance is such that the regulation tends to amplify its effect, an increase in the CLTR will generally decrease the effective overshoot of the CV.
- If an impulse-type disturbance places the CV outside the zone, a short CLTR is used to quickly return the CV to the "safe zone" where there is no risk of instability.

The variation of the MV is nominal when the regulated variable resides within the designated zone of control. This condition may be exploited successfully in a shared or alternating control scheme. In fact, it could be argued that a form of *pseudo-multivariable* control may be achieved by prioritising the variable to regulate according to the error (see Sections 7.5 and 7.7).

The simplest approach involves altering the CLTR linearly between a maximum value CLTR_{max} corresponding to $\varepsilon(n) = 0$ and two minimum values $\text{CLTR}_{\text{minmin}}$ and $\text{CLTR}_{\text{minmax}}$ for $\varepsilon(n) = \varepsilon_{\text{max}} > 0$ and $\varepsilon(n) = \varepsilon_{\text{min}} < 0$,

$$\varepsilon(n) \geq 0: \quad \text{CLTR}(n) = \text{CLTR}_{\text{max}} + \frac{\varepsilon(n)}{\varepsilon_{\text{max}}} \left(\text{CLTR}_{\text{minmax}} - \text{CLTR}_{\text{max}} \right),$$

$$\varepsilon(n) \geq \varepsilon_{max}: \qquad \text{CLTR}(n) = \text{CLTR}_{minmax},$$

$$\varepsilon(n) \leq 0: \qquad \text{CLTR}(n) = \text{CLTR}_{max} + \frac{\varepsilon(n)}{\varepsilon_{min}}\left(\text{CLTR}_{minmin} - \text{CLTR}_{max}\right),$$

$$\varepsilon(n) \leq \varepsilon_{min}: \qquad \text{CLTR}(n) = \text{CLTR}_{minmin}.$$

See Figure 7.1.

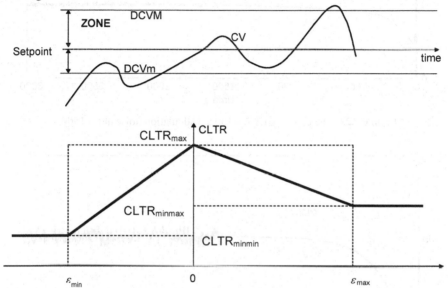

Figure 7.1. Zone control

Consider the example of a second-order process with time constants $T_1 = 30$ s and $T_2 = 60$ s, unity gain and $h = 12$. The system is subjected to a disturbance that may be modelled by a white-noise source passed through a first-order filter of decrement am $= 0.9$ and sample time $T_s = 1$ s. The zone is an even function, *i.e.*,

$$\text{CLTR}(\varepsilon(n)) = \text{CLTR}(-\varepsilon(n)).$$

A $\text{CLTR}_{max} = 150$ s and a set-point of 100 are chosen in conjunction with zone boundaries of $\text{DCVM} = \text{DCVm} = 20$ and a CLTR_{min} of $\text{CLTR}_{min\,min} = \text{CLTR}_{min\,max} = 50$ s. The influence of the zone may be analysed by comparing the CLTR_{eff}, *i.e.*, the measured, effective closed-loop time response and the error types σMV and σCV on the regulated and manipulated variables, respectively.

- For $\text{CLTR}_{max} = 150$ s, $\text{CLTR}_{eff} = 850$ s $\sigma\text{CV} = 2.53$ $\sigma\text{MV} = 0.64$.
- For $\text{CLTR}_{max} = 50$ s, $\text{CLTR}_{eff} = 603$ s, $\sigma\text{CV} = 2.58$ and $\sigma\text{MV} = 1.64$.

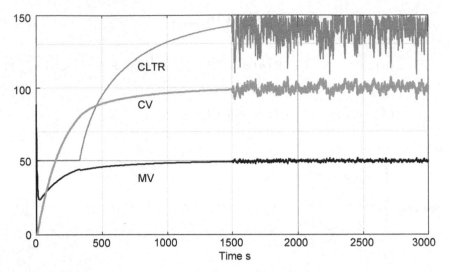

Figure 7.2. Zone control CLTR = 150 s. Calculation noise at t = 1500 s

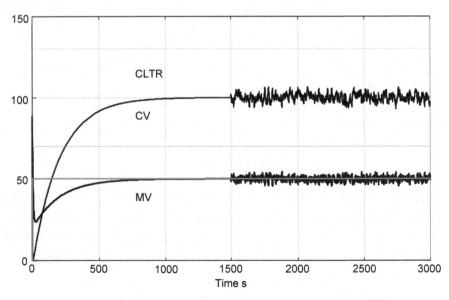

Figure 7.3. Zone control CLTR = 50 s. Calculation noise at t = 1500 s

It can be seen in Figures 7.2 and 7.3 that the effective CLTR is reduced by a factor of 1.4 for approximately the same value of σCV. On the other hand, σMV is increased by a factor of 2.56.

The benefit of this effectively variable CLTR approach is that the result is achieved with a reduction in the risk of instability and controlled variable activity when compared with the direct approach. Note also that there is no perceivable loss in the dynamic performance of the system in the transient phase.

7.3 Cascade Control

Cascade control is also a common strategy. It is characterised by the MV of a predictive regulator (external or master) being used as the set-point for another regulator (internal or slave). The master regulator is used to control the outer loop while the slave regulator maintains the inner loop (see Figure 7.4).

The external regulator is often of level 1 [15] (main process variable, *e.g.*, temperature); whereas the internal regulator is of level 0 (ancillary variable, *e.g.*, flow), as in the classic example of the temperature regulation of a heat exchanger or reactor. This is also the case for mechanical servos, which consist of several nested cascades from the inner to the outer loop, *i.e.*, regulating current, regulating speed and regulating position.

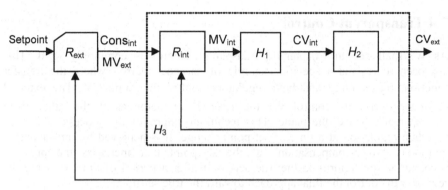

Figure 7.4. Cascade control

The manipulated variable of the external regulator MV_{ext} becomes the set-point of the internal regulator $Cons_{int} = MV_{ext}$.

Consider the case of two predictive regulators in cascade. The internal loop normally incorporates the internal model and actuator constraints. At this point a certain internal CLTR is chosen. If the set-point and the disturbances are such that the system does not reach its operational limits the process loop behaves, to a first approximation, as a first-order system, possibly possessing a pure time delay.

The process loop is controlled by another external predictive regulator, whose internal model is the product of the transfer function H_2 and a first-order, unity gain system (zero position error of the predictive regulator) with time constant T_{ext} . The time constant can be calculated using:

$$T_{ext} = \frac{(\text{internal CLTR})}{3} .$$

It is often the case that the dynamics of the level 0 transfer function H_1 are very fast with respect to H_2 (time constant T_2). In such circumstances an initial approximation of the external regulator's internal model time constant (input

Cons_{int} and output CV_{ext}) of $\text{Tmod}_{\text{int}} = \dfrac{\text{CLTR}_{\text{int}}}{3} + T_2$ is used. This approach facilitates a rapid approximation of the controller. In fact, the normal procedure would consist of identifying the internal process loop H_3 by applying a test signal to its set-point Cons_{int} in order to identify a more precise internal model for the external regulator whose input and output consists of Cons_{int} and CV_{ext}, respectively.

The external loop is adjusted to produce a response that is at least *four times* slower than that of the internal loop. This is consistent with the empirical rule that is usually applied in the case of cascaded systems.

7.4 Transparent Control

As we shall see in Chapter 10, transparent control is commonly used. This approach represents a specific example of cascade control, where the transfer function H_2 is unity and both regulators control the same CV. The external regulator exerts its control via the internal regulator as if the latter were "transparent"; hence, the name. This technique improves the process' ability to maintain regulation at a time of set-point change. It is achieved by introducing a temporary "over-compensation" of the set-point that increases the process' acceleration and returns to the true set-point after a period of time allowing the regulator to control the final approach toward the true set-point.

It is an additive procedure that allows the system stability, controlled by the internal regulator, to be decoupled from the dynamic response for a change in set-point or in response to a disturbance. Traditionally, the aeronautics and robotics industry have made use of this technique.

This technique is commonly used for level control or, more generally, an integrator process. Transparent control is used for the improvement in performance that it provides, but especially for its easy step-by-step implementation. The level of a boiler is a process that may be represented by a transfer function of the form:

$$H(s) = \frac{K}{s(a_1 + sT)} \ .$$

A time delay is frequently introduced because of actuator and sensor latencies. If a PID regulator is used to track the set-point accurately another integrator must be introduced into a loop already containing one integrator, with all the associated difficulty in adjustment that follows. The implementation of level control is critical as manual control is physically difficult; it is imperative neither to let the drum run empty nor overflow.

The proposed solution consists of implementing a proportional regulator with a gain K as an internal controller and to superimpose an external predictive regulator (see Figure 7.5).

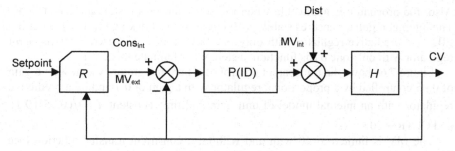

Figure 7.5. Transparent control

The purpose of the regulator P is not to maintain the set-point in the presence of a disturbance, but to stabilise the loop. This will reassure both instrumentation specialists and operators alike. The regulator R eliminates any offset and enhances dynamic performance.

It is important to use a proportional-only regulator P when applying the constraint transfer function. This would be practically impossible with a full PID regulator (as in PI and D included). The constraint transfer function would become purely algebraic:

$$\text{Cons}_{intM} = \text{CV}(n) + \frac{MV_{int}}{K} \ .$$

The Cons_{intM} value behaves like a constraint on the manipulated variable of the external predictive regulator. The procedure is sometimes easier to implement than that of the decomposition principle discussed in Chapter 2 because of its compatibility with conventional theory.

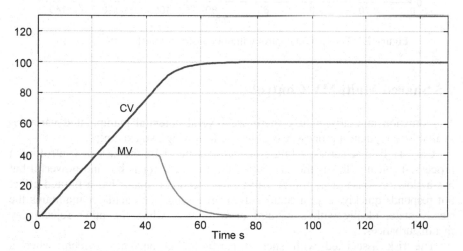

Figure 7.6. Transparent control with constraint transfer $MV_{MAX} = 40$

Also, the procedure is applicable when the set-point, or a disturbance, is a ramp. The internal regulator ensures stability. But, a non-zero tracking error will exist if a PID or a predictive regulator with only one basis function is used. A transparent regulator with only one basis function, a step, will eliminate this error.

Figure 7.6 shows the transparent control of a pure integrator system with a gain of 0.05 controlled by a proportional regulator with a gain of 0.1 and by a predictive regulator with an internal model of unit gain and time constant $\tau = 1/(0.05)(0.1)$ and $CLTR = 20\ s$.

The MV is limited to 40 with and without a constraint transfer function (see Figures 7.6 and 7.7). The benefit of using this constraint transfer function is also demonstrated. This procedure is convenient to use due to its safe step-by-step implementation.

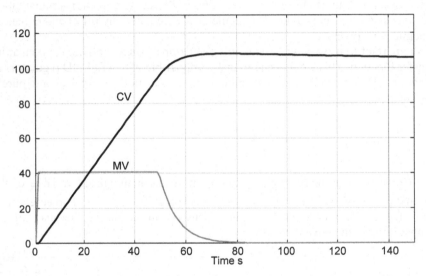

Figure 7.7. Transparent control without constraint transfer $MV_{MAX} = 40$

7.5 Shared Multi-MV Control

The situation may arise where there are several potential manipulated variables available to regulate a process. Consider the following two examples.

Case 1: Two actuators, acting at the same time, share the task of controlling a process. Typically, this is the case when the first actuator is big and powerful but has a slow response (*e.g.*, a large valve). The second actuator has a limited effect but responds quickly (*e.g.*, a small valve). In essence, one actuator maintains the steady state, ideally with a centred or zero MV, while the other rejects any fast disturbances.

The risk associated with such a configuration becomes evident when a disturbance is introduced. The primary actuator will behave as expected and move in the proper direction. But, the secondary actuator will run the risk of moving in the opposite direction, even though the collective sum of their two actions is correct. Thus, not only would the two regulators not cooperate – they would actually oppose each other! This situation arises because there is only one equation to satisfy with two unknowns. Consequently, some unacceptable solutions may result.

Case 2: The situation may also arise where two actuators act in sequence, following the applied signal. A common example is the "hot/cold" split-range control of a thermal process (for example, a chemical reactor). The gains and dynamics of the two actuators are not identical. Steam is injected into the coolant loop to rapidly heat the process; whereas it is necessary to pass the coolant through a heat exchanger to cool the system.

The discrepancies between the actuators' gains may be easily corrected. But, if the dynamics are different, the tuning of the controllers must be different. Under such conditions, it is almost impossible to achieve satisfactory results with a single PID regulator. Indeed, it is not possible to tune the regulator to two different dynamics *a priori* unless it is adjusted to a very "loose" solution. This compromise is frequently encountered in the chemical industry but is less than satisfactory.

In the two cases that follow, the benefit of separating the free and forced outputs will be seen.

7.5.1 Coupled Actuators

Consider the case where two manipulated variables act on the process simultaneously (see Figure 7.8): a fast variable MV_1 that must tend towards 0 in steady state, and a slow variable MV_2 that ensures the steady state.

For simplicity, assume that the overall transfer function is identical for both manipulated variables. This implies that the transfer functions of the actuators must be negligible with respect to that of the process.

The control equation has two free outputs terms and, therefore, two forced output terms:

$$\left(\text{Setpoint} - s_p\left(n\right)\right)l_h = s_{m1}\left(n\right)a_1^h + K_1\left(1 - a_1^h\right)MV_1\left(n\right) - s_{m1}\left(n\right)$$
$$+ \ldots + s_{m2}\left(n\right)a_2^h + K_2\left(1 - a_2^h\right)MV_2\left(n\right) - s_{m2}\left(n\right) .$$

There is only one degree of freedom available to solve the equation. One approach is to search for a solution that minimises the cost function C that is composed of the terms $C = C_1 + C_2 + C_3$. Assume,

$$C_1 = \left[Z - K_1\left(1 - a_1^h\right)MV_1\left(n\right) - K_2\left(1 - a_2^h\right)MV_2\left(n\right)\right]^2$$

are used to satisfy the reference trajectory:

$$C_2 = q_1 MV_1\left(n\right)^2 .$$

Decreasing MV_1 toward 0 gives:

$$C_3 = q_2 \left[MV_2(n) - MV_2(n-1) \right]^2$$

Thus, reducing the variations of MV_2. Minimisation of the function $C = C_1 + C_2 + C_3$, in a least-square sense, results in solutions for MV_1 and MV_2 of the control equation, where q_1 and q_2 are two further explicit weighting parameters.

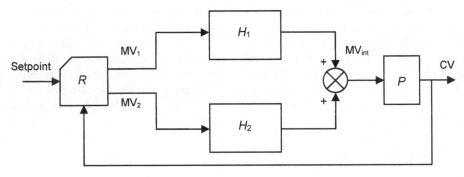

Figure 7.8. Shared multi-MV control

Defining C in terms of MV_1 and MV_2 and setting the partial derivatives to zero results in the following two equations that provide optimal values for $MV_1(n)$ and $MV_2(n)$:

$$\left[Z - U_1 MV_1(n) - U_2 MV_2(n) \right] U_2 - q_2 \left(MV_2(n) - MV_2(n-1) \right) = 0$$

where $U_1 = K_1 (1 - a_1^h)$ and $U_2 = K_2 (1 - a_2^h)$. The parameters q_1 and q_2 must be tuned during simulation. This procedure is interesting because it demonstrates the connection between the theoretical minimisation of the various criteria placed on the desired characteristics. On the other hand, the "*tuned during simulation*" concept is not acceptable to the industrial practitioner because it is not consistent with the spirit of our approach.

Another strategy consists of searching for the inevitable connection that exists between the fast manipulated variable MV_1 and the slow manipulated variable MV_2.

By eliminating the term between the brackets $\left[Z - U_1 MV_1(n) - U_2 MV_2(n) \right]$, an equation representing MV_1 in terms of MV_2 may be found:

$$\mathrm{MV}_1\left(n\right)=\frac{q_2}{q_1}\frac{U_1}{U_2}\left(\mathrm{MV}_2\left(n\right)-\mathrm{MV}_2\left(n-1\right)\right),$$

where MV_1 appears to within a constant times of the derivative of MV_2.

This solution is illustrated in Figure 7.9. A regulator R, which controls the physical output of the process, is introduced. The process is subjected to a slow and fast actuator that produces a "slow manipulated variable" of value MV_2 that controls the slow actuator, and a "fast manipulated variable" MV_1, generated by differentiating the slow manipulated variable using a high-pass filter (HPF) that controls the fast actuator. The two models (slow actuator × process) and (fast actuator × high-pass filter × process) have free and forced outputs that are introduced into the control equation. This results in a manipulated variable that not only acts directly on the slow process but also on the fast process via a high-pass filter.

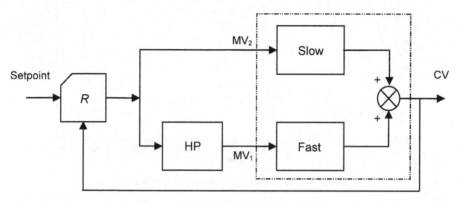

Figure 7.9. Shared control (slow/fast)

The complete system behaves as a single process and the regulation reduces to a classical control problem where the usual parameters (CLTR) are determined. The duration of the transient behaviour of the fast MV is determined by the time constant of the high-pass filter:

$$\mathrm{MV}_1\left(n\right)=\mathrm{MV}_1\left(n-1\right)a_f+b_f K\left(\mathrm{MV}_2\left(n\right)-\mathrm{MV}_2\left(n-1\right)\right).$$

The relationship between the amplitudes of the slow and fast MVs is fixed for $n=1$, by the product term $b_f K$. For the first value of $\mathrm{MV}\left(n=1\right)$ we have $\mathrm{MV}_1\left(1\right)=\mathrm{MV}_2\left(1\right)b_f K$, which may be physically interpreted directly. From which we may deduce that $a_f=\mathrm{e}^{-\frac{T_s}{T_{\mathrm{filter}}}}$ and $b_f=1-a_f$.

As an example, consider the case of a third-order system that is composed of a slow process and a filtered fast process.

- Slow process: $K_1 = 1$, $\tau_1 = 70$ s .
- Fast process: $K_r = 1$, $\tau_r = 20$ s .
- High-pass filter: $G_f = 20$, $\tau_f = 15$ s .
- CLTR $= 150$ s .
- Slow process speed constraint: $dMV = 2$.
- Fast process amplitude constraint: $MV_{max} = 25$.

The slow MV is in speed saturation and the fast MV touches its maximum value. The CV behaves correctly (see Figure 7.10).

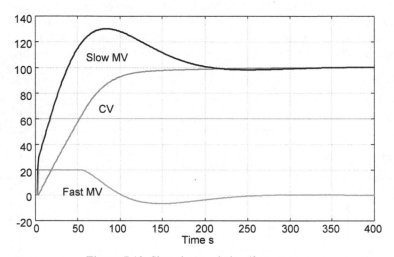

Figure 7.10. Shared control-slow/fast actuators

7.5.2 Split-range Control

The two regulators that are present in this configuration operate in an alternating fashion, *i.e.,* only one manipulated variable is active at any particular instant as it forces the other MV to zero, *e.g.,* for operational reasons, a valve is never completely closed, in order to prevent blockage or "sticky valve" (stiction). The transfer functions can be very different and the two regulators R_1 and R_2 (hot and cold) are adjusted independently. A problem arises in that, for a given combination of set-point and regulated variable, the two proposed actions produced by the two regulators are not known *a priori*.

A decision mechanism, *i.e.,* a supervisor, applies the most appropriate manipulated variable and informs the regulators of the choice made by supplying the internal models of the regulators with the chosen value (see Figure 7.11). At the time of switching from one actuator to the other, one MV is set to 0, but the free output (*i.e.,* the term of the MV that represents the influence of the past) is non-zero, and must be taken into account in the control equation of the other regulator. The control equations have a very particular structure, with two free outputs. We consider the simple case of two first-order processes with $h = 1$.

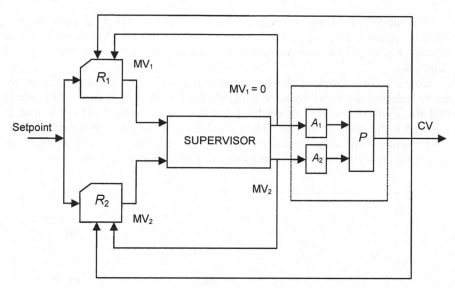

Figure 7.11. Shared control of two actuators $\mathrm{MV_2}$ active, $\mathrm{MV_1} = 0$

Regulator R_1:

$$\left(\text{Setpoint}(n) - s_\mathrm{p}(n)\right)(1 - \lambda_1) = \mathrm{MV_1}(n)\mathrm{K}_\mathrm{m1}b_\mathrm{m1} + s_\mathrm{m1}(n)a_\mathrm{m1} - s_\mathrm{m1}(n)$$
$$+ s_\mathrm{m2}(n)a_\mathrm{m2} - s_\mathrm{m2}(n) \; .$$

In the case where $h = 1$, $b_\mathrm{m1} = 1 - a_\mathrm{m1}$ and $b_\mathrm{m2} = 1 - a_\mathrm{m2}$, we obtain:

$$\left(\text{Setpoint}(n) - s_\mathrm{p}(n)\right)(1 - \lambda_1) = \mathrm{MV_1}(n)\mathrm{K}_\mathrm{m1}b_\mathrm{m1} - s_\mathrm{m1}(n)b_\mathrm{m1}$$
$$- s_\mathrm{m2}(n)b_\mathrm{m2} \; .$$

Regulator R_2:

$$\left(\text{Setpoint}(n) - s_\mathrm{p}(n)\right)(1 - \lambda_2) = \mathrm{MV_2}(n)\mathrm{K}_\mathrm{m2}b_\mathrm{m2}$$
$$- s_\mathrm{m2}(n)b_\mathrm{m2} - s_\mathrm{m1}(n)b_\mathrm{m1} \; .$$

In general, the transfer functions of both regulators are different, which gives rise to two different CLTRs being required. The final stage consists of selecting the more appropriate MV to guide the system having calculated the two potential MVs.

The control equations may be determined algebraically. The gains K_m1 and K_m2 of the two processes A_1P and A_2P are of opposite sign and each regulator may propose a negative MV that has no physical interpretation.

The supervisor's logic is simple: if the value of one MV is positive and the other is negative, the positive MV is retained and the negative MV is set to 0. The internal models of the two regulators are supplied by the MV selected by the supervisor.

Consider the example of driving a car. It is possible to slow down the car by either reducing the petrol intake or applying the brakes. In this case, it is possible that the signs of the two manipulated variables are identical. The MV that is selected is the same as the one that was selected at the previous sampling time. Under normal operating conditions, a switching arrangement will result in an "acceleration – breaking" behaviour using two conventional regulators.

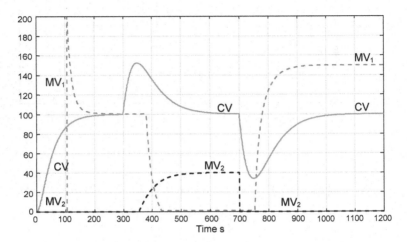

Figure 7.12. Split-range control

Consider the example of a process with unity gain and a time constant $T_m = 50$ s controlled by two actuator transfer functions $A_1 = \dfrac{1}{1+s25}$ and $A_2 = \dfrac{-0.5}{1+s10}$. The coincidence point is $h = 40$ and the closed-loop time response of the two regulators is CLTR $= 120$ s .

When a set-point of 100 is applied, the regulator generates an overshoot on MV_1, while $MV_2 = 0$. At 300 s, an additive disturbance acts on the process. To counteract this disturbance, MV_1 decreases and becomes negative, while MV_2 appears to oppose the effect of the disturbance. At 700 s, the disturbance becomes positive and an inverse response occurs. At this point MV_2 returns to 0 and MV_1 moves to counter the disturbance (see Figure 7.12).

It is possible to take the cross-over effects of the free outputs into consideration by analysing the physical geometry of the actuators. The implementation is straightforward as each regulator is adjusted in an automated manner without compromising the regulation of the other.

7.6 Estimator

A problem that is frequently encountered, not only from a control perspective, is to determine an internal process variable of known structure, e.g. identifying the level of exothermic reaction present in a chemical reactor. This problem is equivalent to using sensors to estimate a signal, by means of a mathematical model, in real time using the measured sensor outputs. An algebraic solution may be used assuming that the estimation is stationary.

In certain circumstances the coefficients may be identified – through the use of historical tables, for example. But, as the required signal represents the input to a dynamic process, it would be necessary to invert the dynamic transfer function. The same input/output relationship is always used:

$$S = HE,$$

where E is the input, S the output, and H the transfer function output of a mathematical model.

- *Simulation*:   To find S, knowing H and E.
- *Identification*:   To find H, knowing E and S.
- *Control*:   To find E, knowing H and S (desired output).
- *Estimation*:   To find E, knowing H and S (actual output).

Control and estimation requires the inversion of the transfer function H: $E = H^{-1}S$.

The strategy adopted is to use a regulator in the role of an inverter and to identify the process input that generates the measured output. It is a classical problem in automation known as *observation* (or *reconstruction*).

However, such solutions do not always exist. For instance, consider the case where two identical signals, separated by an unknown shift in time, are input to a process possessing a pure time delay. It is not possible to distinguish between the individual input signal responses solely by observing the process output.

In the non-delayed case, the strategy consists of taking a theoretical servosystem where the simulated process P and the internal model of the regulator are identical $M \equiv P$. The internal model is controlled by a regulator R_2 whose set-point is the measured output of the process (see Figure 7.13).

Since the control of this regulator is "perfect", the output of the process $s_p(n)^*$ is identical to the measured output $s_p(n)$ and, in accordance with the conditions stated earlier, the inputs $MV_1(n) + \text{Pert}(n)$ and $MV_2(n)$ are identical, i.e., $\text{Pert}(n) = MV_2(n) - MV_1(n)$.

The regulator R_2 may be "tightly" tuned since the model and process are identical; thus, there are no potential problems with robustness. However, the measured value of $s_p(n)$ contains noise B (see Figure 7.13) and the estimator amplifies any noise present at the input of a low-pass system resulting in a potentially large amplitude. The effect of such noise on the reconstructed input signal may be minimised by adapting the CLTR of the reconstruction controller.

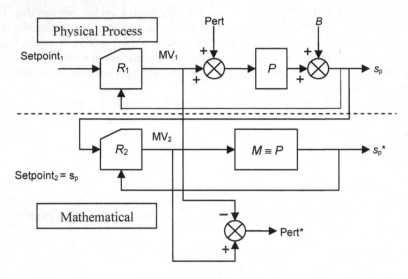

Figure 7.13. Estimator

Consider the common example of a first-order system where $K = 1$, $\tau = 30$, delay $= 20$ and CLTR $= 70$ s subject to a set-point setpoint $= 100$ and an unknown disturbance Pert acting at the process input.

Industrial processes, where potentially disastrous disturbances are a possibility, would benefit most from this procedure. The injection of reactant into a batch reactor vessel, which is common in the chemical and pharmaceutical industries, represents a typical example of such a process. In such circumstances, the severity of the resulting exothermic reaction may be such as to endanger the operation of the plant.

The response of the jacket-temperature control system may be improved by using a real time estimate of the level of exothermic reaction as a measured, feedforward disturbance. Two disturbance classifications will now be considered; namely state and structural disturbances.

7.6.1 State and Structural Disturbances

As mentioned previously, it is not possible to determine in real time whether a disturbance is due to model mismatch (structural parameter) or is the result of an external disturbance signal (state variable). Consider the case where the process gain K_p and model gain K_m are different. In steady state, the process output S_p may be expressed as:

$$S_p = K_p \left(MV_1 + Pert \right) = Setpoint_1 .$$

If we assume that the process is stable, the steady state error is zero. Thus,

$$S_p^* = K_m MV_2 = \text{Setpoint}_2 = \text{Setpoint}_1 .$$

The estimated disturbance is given by: $\text{Pert}^* + MV_1 = MV_2$. Eliminating MV_1 and MV_2 we get:

$$\text{Pert}^* = \text{Pert} + \text{Setpoint}\left(\frac{1}{K_m} - \frac{1}{K_p}\right),$$

where $\text{Setpoint} = \text{Setpoint}_1 = \text{Setpoint}_2$.   This   remarkable   result   clearly demonstrates the combined effect of the state and structural errors!

   When $K_p > K_m$ this results in $\text{Pert}^* > \text{Pert}$, $i.e.$, the estimated disturbance is too large. This over-estimation of Pert^* compensates for the under-estimated model gain K_m, which is too low. But, the overall product $\text{Pert}^* K_m$ is closer to $\text{Pert} K_p$. However, when the controller is in quasi-steady state, it is sometimes possible to go further. We have two unknowns K_p and Pert. If both parameters are assumed to be constant, it is possible to estimate them using a least-squares or a recursive prediction-error approach. We use two different set-points that result in two different steady states. The estimated disturbances for each steady state are given by:

$$\text{Pert}_1^* = \text{Pert} + \text{Setpoint}_1\left(\frac{1}{K_m} - \frac{1}{K_p}\right)$$

and

$$\text{Pert}_2^* = \text{Pert} + \text{Setpoint}_2\left(\frac{1}{K_m} - \frac{1}{K_p}\right).$$

Combining these equations it is possible to extract K_p using:

$$K_p = \frac{K_m}{1 - K_m \dfrac{\left(\text{Pert}_1^* - \text{Pert}_2^*\right)}{\text{Setpoint}_1 - \text{Setpoint}_2}}.$$

Substituting for K_p in Pert_1^* and Pert_2^* above and eliminating K_m gives:

$$\text{Pert}^* = \text{Pert}_2^* . \text{Setpoint}_1 - \frac{\text{Pert}_1^* . \text{Setpoint}_2}{\text{Setpoint}_1 - \text{Setpoint}_2}.$$

Thus, knowing K_p and Pert^* the model gain may be determined.

7.6.2 Feedforward Variable

It is also possible to use the estimated input disturbance as a feedforward variable. But this estimation is just a mathematical operator whose input is the process output and whose output is to be fed back to MV_1 by some means. So, from an input–output perspective, we just add another feedback loop around the main controller. Thus, this additional loop may affect the global stability of the closed-loop system. In practice, an attenuated version of the estimated disturbance is fed back in the form: $Dist^{**} = K.Dist^*$, where $K < 1$.

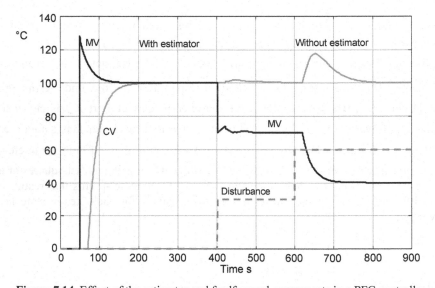

Figure 7.14. Effect of the estimator and feedforward components in a PFC controller

Example 7.1. Consider Figure 7.14, at time 400 s a step disturbance of amplitude 30 is introduced. The estimator and feedforward components are active at this point. At 600 s, another step disturbance is introduced but, at this point, the estimator and feedforward components are deactivated.

The results demonstrate that there is a reduction in the error response by a factor of 10 when the estimator and feedforward components are active.

Programming the observer has the effect of increasing the disturbance rejection capacity of the controller. However, if the noise level increases then the CLTR of controller R_2 should be increased. This feedforward component should be preferably deactivated when changing set-point.

Example 7.2. Figure 7.15 illustrates the temperature profile resulting from the distillation of acetone solvent in a batch reactor (boiling temperature = 56 °C). The jacket temperature is controlled using a PFC controller with an initial set-point of 60 °C, which is later adjusted to 70 °C. The model and the mass temperature remain similar until 200 s (temperature = 47 °C). At this point the model and process responses begin to diverge as the solvent begins to evaporate. At 400 s, the jacket temperature is increased. This produces an increase in evaporation while the mass temperature tends towards its boiling point. The real time estimate of the evaporation disturbances tracks the difference between the mass and model temperatures quite well. In this case the accuracy of the estimator is easy to verify.

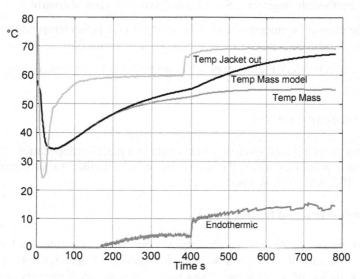

Figure 7.15. On-line estimation of the endothermic reaction during the distillation of acetone solvent

7.6.3 Calibration

The classical approach to evaluating the exothermicity or endothermicity of the chemical reaction in the reactor involves measuring the jacket input and output temperatures and computing the energy extracted from or fed into the reactor during the batch time. The net energy transfer is given by:

$$W_1 = \sum Q(n)^* \left(\text{Tempout}(n) - \text{Tempin}(n) \right) \quad \text{in kWh,}$$

where Q, Tempout and Tempin represent the jacket flow, output and input jacket temperatures, respectively.

$$W_2 = k \sum \text{Dist}(n) \text{ should be set equal to } W_1 .$$

Computing $W_2 = W_1$ permits the evaluation of the value of k. Therefore, it is now possible to evaluate in real time the instantaneous exo- and endothermicity.

It is to be noted that implementing this observer is straightforward since it duplicates the real time controller scheme where the physical process is replaced by its known model and the CLTR of R_2 is much smaller, with no instability issues since it is a pure mathematical structure.

Thus, this procedure may be implemented with minimum effort and is routinely incorporated into modern PLC controllers. Moreover, some chemical and pharmaceutical producers tend now to act on the reactant flow to achieve a specified exothermic target rate. So, we now have two types of control:

- the classical temperature controller acting on the jacket temperature;
- the exothermic rate controller acting on the reactant flow.

The interaction between the two controllers is implicitly taken into account.

7.7 Non-linear Control

There is no general theory governing the simple and practical application of control to non-linear processes. Thus, the discussion will be limited to the treatment of some specific, but common, cases.

7.7.1 Non-linearities of the MV or CV

It is first necessary to distinguish between functional variables and physical variables. The variable MVF is a *functional variable* that operates linearly on the dynamic equation of the process. A non-linear relationship exists between MVF and the variable MVΦ that physically acts on the process. If this relationship is one-to-one mapping (*i.e.*, a unique mapping exists between each value of MVF and its corresponding MVΦ and *vice versa*) and stationary, it is possible to determine a regulator with MVF and subsequently calculate the appropriate value of MVΦ to apply (see Figure 7.16).

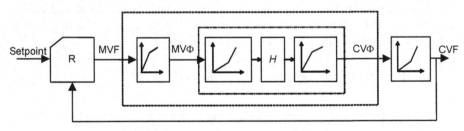

Figure 7.16. "Input-output" non-linearities

The CV should be treated in the same manner. If a one-to-one mapping and stationary relationship exists between the CVΦ and CVF, the reference trajectory may be initialised using a calculated point CVF according to the measured CVΦ.

Consider the case of a first-order dynamic process, where the relationship between MVF and MVΦ is generally a polynomial of degree B, where $MVΦ = MVF^B$. The output is a polynomial of degree A $CVΦ = k.CVF^A$. This simplification facilitates the inversion of these functions. In general, it is possible to represent these non-linear characteristics as a series of piece-wise approximations.

Since the internal model of the regulator is linear, it is possible to transform the set-point value Setpoint and the initial reference trajectory point CV^* using the inverse transformation:

$$\text{Setpoint*} = \text{Setpoint}^{(1/A)}, \quad CV^* = CV\Phi(n)^{(1/A)} \ .$$

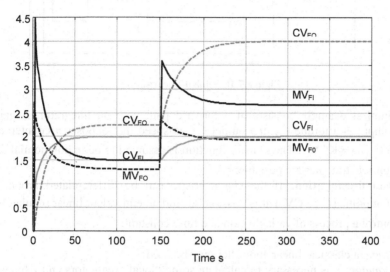

Figure 7.17. Stationary, non-linear features of MV and CV

The value of MVF may be calculated in the usual fashion to supply the internal model. The manipulated variable, which is to be physically applied to the process, may then be determined by using the inverse function to calculate the corresponding value of $MV\Phi$, *i.e.*, $MV\Phi(n) = MVF(n)^{(1/B)}$.

Figure 7.17 demonstrates the results of the application of this procedure for $A = 0.5$, $B = 1.5$, a process gain $k_p = 1.5$, a model gain $k_m = 1$, time constant $T = 40$ s, CLTR $= 100$ s with a physical Setpoint $= 2$ and an additive disturbance on the output at 150 s.

7.7.2 Non-linear Stationary Feedback

This particular case is relatively common and may be generalised. Energy transducers, such as combustion engines or electric motors, receive energy in one form and generate a mechanical torque. This torque typically opposes the frictional and other counteracting forces that, in general, are highly non-linear (see Figure 7.18).

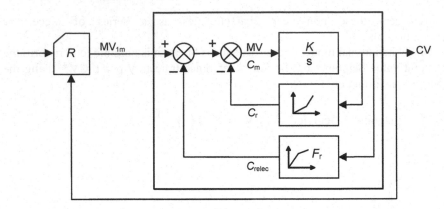

Figure 7.18. Global linearisation

The simplest example consists of a forward path with an integrator of gain K whose input is a torque C_m; the output CV is speed that opposes a counteracting torque $C_r(n) = KCV^A$. The transfer function is assumed to be in polynomial form. Two control strategies are possible:

Local linearisation: If the characteristic can be differentiated around the operating point (C_{m0}, CV_0) it is possible to find an equivalent linear regulator that will control a process of variable gain and time constant.

Advantage: a classical linear controller may be used.
Disadvantage: it is necessary to calculate some "local" regulators and tabulate all tuning parameters. As with all look-up table techniques, this method suffers from a lack of flexibility when the characteristics vary.

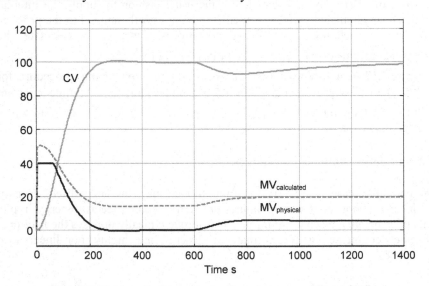

Figure 7.19. Linearisation control with stationary, non-linear feedback

Global linearisation: Conceptually, this procedure mixes physical variables with electrical functional variables. If the electrical CV is fed back via a transfer function having a stationary characteristic R, complementary function $C_{relec}(n) = R[CV(n)]$, then the opposing torque becomes apparent. $F_1(x)$ and $F_2(x)$ are complementary, so $F_1(x) + F_2(x) = Kx$.

$C_r^*(n) = C_{relec}(n) + C_r(n)$ is   assumed   linear.   Subsequently,   setting $C_r^*(n) = k_r CV(n)$, where k_r is an arbitrary constant results in a characteristic R that is given by: $R[(CV)] = k_r CV(n) - KCV^A$.

The feedback is assumed globally linear and the controller transfer function $H(s)$ possesses a gain and constant dynamics:

$$H(s) = \frac{1/k_r}{1 + s\tau}, \text{ with } \tau = \frac{1}{k.k_r} .$$

Choosing CLTR equal to $CLTR_0$ results in a time constant $\tau = \dfrac{CLTR_0}{3}$ where

$k_r = \dfrac{3}{CLTR_0 k}$. A classical PFC controller may be used with constant features. The physical constraint of maximum torque C_{max} is a constant value that must be transferred. Under these conditions, the maximum constraint C_{max}^* of the MV functional MV_{linear} is no longer constant but is instead a function of the speed CV. The non-linear characteristic $C_r = F[CV(n)]$ represents curves of "constant energy".

$$CV(n) = CV0 = \text{constant}, \quad C_{m0} - C_r(CV_0) = 0 .$$

Thus, the complementary function R has a simple physical interpretation.

Experimentally, this procedure was implemented on a process $F = CV(n)^2$ with a process gain $k_p = 0.01$ and a corresponding model gain $G = 1.2/k_r$ compared with the theoretical value $1/k_r$. A $CLTR = 100$ s was chosen and an additive disturbance was applied at 600 s. Figure 7.19 illustrates the resulting behaviour of the system.

7.8 Scenario Method

PFC control is simple because, as a consequence of the linear superposition theorem, it is possible to isolate the free and forced outputs and calculate the control incrementally. However, in the case of non-linear systems, this approach is not generally feasible.

If the process state is known, by measurement or estimation, it is possible to create a *scenario* with the aid of the model. Using an appropriate method with a non-linear solver, the MV scenario describing the evolution of the model output toward the specified point on the reference trajectory may be calculated. Several techniques may be used.

In the case where the process doesn't possess any discontinuities, *i.e.*, if it is locally differentiable without any singularities, a local projection may be utilised that would change at each instant.

The simplest approach consists of creating two projections from the known state at the current instant n. Assume, for example, $MV(n+i) = MV(n) + dMV$, where dMV is assumed to be a fraction of $MV(n)$, *e.g.*, $dMV = 0.05MV(n)$, the model behaviour is simulated producing an output $S_1 = s_m(n+i)$. The scenario is then re-calculated with another value of $MV(n+i)$: $MV(n+i) = MV(n) - dMV$ yielding an output S_2. Linear extrapolation or interpolation may then be used to determine $MV(n+1)$ using the values for S_1 and S_2 resulting from $dMV = -0.05MV(n)$ and $dMV = +0.05MV(n)$. This solution results in the convergence of the process model output on the specified coincidence point.

A bias results as a consequence of using a realigned model. This is particularly true in this case as the solution method adopted here is primitive. However, the solution may be improved by using a classical method for solving non-linear algebraic equations such as the Raphson – Newton method.

It would be beneficial to combine this result with a transparent control configuration. This would have the dual effect of reducing the bias and improving the accuracy of the control equation.

Consider the simplified case of a biological photosynthesis reactor where the manipulated variable is the degree of incident light flux and the regulated variable corresponds to the concentration of biomass. A simplified model of this system is given by:

$$CV(n) = CV(n-1)p_1 + \frac{p_2 MV(n-1)}{1 + kMV(n-1)}(C_0 - CV(n-1)),$$

where $p_1 = 0.9$ and $p_2 = 0.01$, and k and C_0 are the identified parameters $(k = 0.1, C_0 = 50)$.

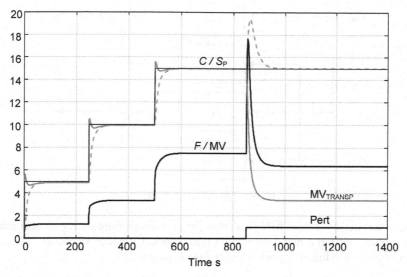

Figure 7.20. Scenario control of a non-linear process

A transparent control configuration is used (see Figure 7.7) with an external regulator consisting of a first-order model with a time constant $T = \dfrac{\text{CLTR}_{int}}{3}$. The CLTR of the internal regulator loop CLTR_{int} is determined by experimentation. A satisfactory behaviour for a change in set-point and disturbance rejection results, without the use of an iterative optimisation technique, when a value of $dMV = 0.4$ is selected (see Figure 7.20).

7.9 2MV/2CV Control

It is not intended that the 2MV/2CV technique should compete with the multivariable approach commonly encountered in the petrochemical industry. Several types of production unit in this industry are based on the binary distillation column. Typically, such units are controlled by manipulating the rate of reflux and the flow rate of reheating fluid at the column base.

A two-dimensional, multivariable controller implemented in block form on simple PLCs would conveniently extend the range of basic regulators and offer benefits to other industrial sectors.

7.9.1 The Problem Being Addressed

In general, taking manipulated variable constraints into consideration constitutes a complex problem even if the control itself does not present any particular problems.

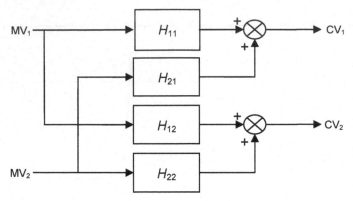

Figure 7.21. 2MV/2CV process

The controller constraint MV field may be visualised more easily by representing the problem within a two-dimensional framework. Consider a process that may be described by the following relationship (see Figure 7.21)

$$CV_1 = H_{11}MV_1 + H_{21}MV_2$$
$$CV_2 = H_{12}MV_1 + H_{22}MV_2 \ .$$

Assume that the elementary transfer function H_{ij} is asymptotically stable with gains K_{ij}. The necessary, but insufficient, condition $D = K_{11}K_{22} - K_{12}K_{21} \neq 0$ must be satisfied to ensure the steady state requirement.

If one of the transfer functions is zero, *e.g.*, $H_{21} = 0$, the control amounts to regulating CV_1 by MV_1 and compensating for the effects of MV_1 on the control of CV_2 using MV_2. Thus, a full, non-triangular, transfer function matrix $H(ij)$ is required. The processes are of an appropriate order and considered here without the presence of a pure time delay, for simplicity.

Remark 7.1. A particular methodology for controlling a simple but practical case is presented here and, as such, is not intended as a general approach for regulating all processes.

7.9.2 Control Computation

The following procedure demonstrates the steps required to determine a controller for SISO processes. Consider a simplified example where there is only one coincidence point h on the two reference trajectories that affect the outputs CV_1 and CV_2. The CLTRs may be different. Assume the models:

$$CV_2(n) = s_{m12}(n, MV_1(n)) + s_{m22}(n, MV_2(n)) \ .$$

Only one basis function is used, *i.e.*, the step. We note that the free and forced unit outputs of the models sm_{ij} are given by S_{Lij} and S_{Fij} and the set-point is referred to as Setpoint$_i$, where $l_{hi} = 1 - e^{-3T_s/\text{TRBF}_i}$. The resulting control equations are given by:

$$\left[\text{Setpoint}_2 - CV_2(n)\right]l_{h2} = S_{L12}(n) + MV_1(n)S_{F12}(h) - s_{m12}(n)$$
$$+ S_{L22}(n) + MV_2(n)S_{F22}(h) - s_{m22}(n)$$

and

$$\left[\text{Setpoint}_1 - CV_1(n)\right]l_{h1} = S_{L11}(n) + MV_1(n)S_{F11}(h) - s_{m11}(n)$$
$$+ S_{L21}(n) + MV_2(n)S_{F21}(h) - s_{m21}(n)$$

Assuming,

$$Y_1(n) = \left[\text{Setpoint}_1 - CV_1(n)\right]l_{h1} - S_{L11}(n) + s_{m11}(n) - S_{L21}(n) + s_{m21}(n)$$
$$Y_2(n) = \left[\text{Setpoint}_2 - CV_2(n)\right]l_{h2} - S_{L12}(n) + s_{m12}(n) - S_{L22}(n) + s_{m22}(n)$$

the following system of linear equations must be solved:

$$Y_1(n) = MV_1(n)S_{F11}(h) + MV_2(n)S_{F21}(h)$$
$$Y_2(n) = MV_1(n)S_{F12}(h) + MV_2(n)S_{F22}(h)$$

A solution exists if : $D_h = S_{F11}(h)S_{F22}(h) - S_{F12}(h)S_{F21}(h) \neq 0$. Thus, D_h depends explicitly on the choice of h that can be calculated *a priori* and maximized. The control solutions $MV_1(n)*$ and $MV_2(n)*$ may then be determined.

7.9.3 Constraints on the MVs

The constraints field is reduced to a rectangular domain in the (MV_1, MV_2) space. The extreme constraints values for each MV are taken (see Figure 7.22).

If the co-ordinate point *M*, where $M = \left[MV_1(n) \ MV_2(n)\right]$, is within the rectangle of constraints, the control is applied. The forced outputs S_{Fij} are never zero. Consequently, D_1 and D_2 are not parallel to the axes and so cut the sides of the constraints' rectangle. The cost function *C* is defined by:

$$C(MV_1, MV_2) = \lambda\left((Y_1(n) - MV_1(n)S_{F11}(h) - MV_2(n)S_{F21}(h)\right)^2$$
$$+ (1 - \lambda)\left(Y_2(n) - MV_1(n)S_{F12}(h) - MV_2(n)S_{F22}(h)\right)^2, \text{ where } 0 \leq \lambda \leq 1.$$

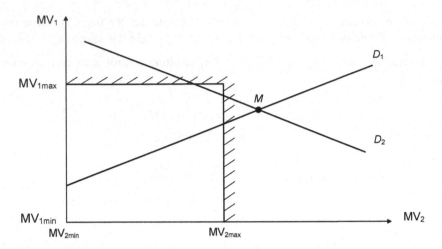

Figure 7.22. 2MV/2CV constraints on the MVs

If priority is to be placed on the CV_1 specification, λ will be chosen as $\lambda = 1$. On the other hand, $\lambda = 0$ is chosen if the CV_2 specification is to be prioritised. All intermediate values between 0 and 1 result in a compromise.

Assuming the $MV*$ solution becomes saturated, *e.g.*, $MV_1 = MV_{1max}$ its value in C will be fixed. Consequently, C becomes a function dependent solely on one variable in this case MV_2. The value of MV_{2opt} that minimises $C\left(MV_{1max}, MV_{2opt}\right)$ is determined by solving the equation $\dfrac{dC\left(MV_2\right)}{dMV_2} = 0$. Thus,

$$\lambda\left[Y_1(n) - MV_{1max}S_{F11}(h) - MV_2(n)S_{F21}(h)\right]S_{F21}(h)$$
$$+ (1-\lambda)\left[Y_2(n) - MV_{1max}S_{F12}(h) - MV_2(n)S_{F22}(h)\right]S_{F22}(h) = 0 .$$

If we assume,

$$Z = \lambda\left[Y_1(n) - MV_{1max}S_{F11}(h)\right]S_{F21}(h)$$
$$+ (1-\lambda)\left[Y_2(n) - MV_{1max}S_{F12}(h)\right]S_{F22}(h) ,$$

MV_{2opt} is given by:

$$MV_{2opt} = \frac{Z}{\lambda S_{F21}(h)^2 + (1-\lambda)S_{F22}(h)^2} .$$

So, if MV_{2opt} lies between MV_{2min} and MV_{2max}, MV_{2opt} is applied; otherwise, the constrained value will be used. For example, if $MV_{2opt} > MV_{2max}$, then $MV_2(n) = MV_{2max}$ will be used (see Figure 7.22).

In the case where both MV solutions of the control equations are saturated, the MV that was saturated at the previous instant is selected. Figure 7.23 illustrates the case of a process where the transfer functions are:

$$H_{11}(s) = \frac{1}{(1+s20)(1+s30)}, \qquad H_{21}(s) = \frac{0.5}{1+s30},$$

$$H_{12}(s) = \frac{0.9}{1+s60}, \qquad H_{22}(s) = \frac{0.9}{(1+s25)(1+s50)}.$$

The CLTRs are identical: $CLTR = 60$ s   with $h = 80$. The additive step disturbances on the outputs saturate the control. The difference in behaviour for $\lambda = 1$ (CV_1 prioritised, see Figure 7.23), $\lambda = 0$ (CV_2 prioritised, see Figure 7.24) or the compromise $\lambda = 0.5$ (see Figure 7.25) may be observed.

7.9.4 Tuning

Simulation is the preferred technique when the system loop is of a sufficiently high order to render an analytic approach difficult. This involves generating the step response for several values of h using the desired CLTR. These responses are then evaluated based on some cost criterion that is a function of the MV and CV.

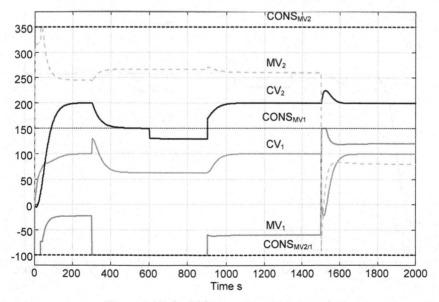

Figure 7.23. $2 \times 2 / \lambda = 1$ multivariable control

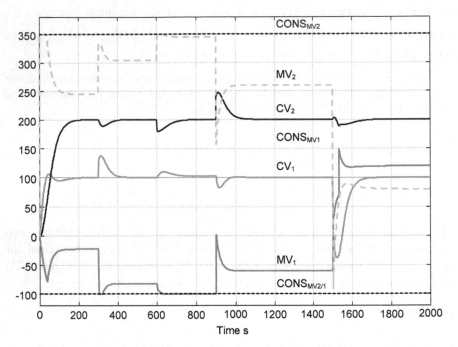

Figure 7.24. $2 \times 2 / \lambda = 0$ multivariable control

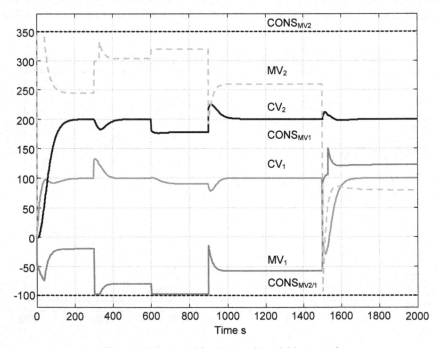

Figure 7.25. $2 \times 2 / \lambda = 0.5$ multivariable control

For example, the following criteria may be considered: steady state stability, local dynamic gain, effective CLTR, maximum values of CV_1, CV_2, MV_1, MV_2 and the average of the absolute error values $(\text{Setpoint} - CV_i)$ at the end of the horizon (*i.e.* the equivalent of a stability criterion). A tuning setting is then selected that is a compromise between all of these values.

7.10 Summary

- PFC is an open technology where numerous control configurations are possible: zone control, cascade control, tranparent control, shared multi-MV control, coupled actuators, split-range control, 2MV/2CV control, *etc*.
- An error estimator coming from either a model mismatch or an additive disturbance produces a correction signal that may be used as a feedforward variable. The detection of an exothermic reaction in batch-reactor control is a practical application of this procedure.

5.10 Summary

8

Parametric Control

Abstract. In this chapter we consider parametric control. This is a novel approach that is particularly useful in the control of thermal processes. The regulation of the hot/cold fluid flow in a batch reactor would represent a typical application domain for this technique. Parametric control is unusual in that it acts on a *structural parameter,* as opposed to classical control that acts on a *state input.* The technique uses a realigned model approach. Thus, to remove any possible bias, the parametric controller is cascaded with a classical PFC controller, that uses an independent internal model implementation. Since passive thermal exchange processes can be represented by linear systems with real poles, the stability of the controller is ensured.

Keywords: Parametric control, parametric instability, heat exchanger, back calculation, enthalpy control

8.1 Parametric Instability

All the problems considered thus far have been solved using regulators that act on *variables* of the process state, input and *MV* of the process. Typically, the process parameters are constant but they may also vary with time, *e.g.*, the input load of a unit. However, processes exist where the controlling action may not be referred to as an "MV" since this action operates on a *parameter* of the process and not on a state variable. As we shall see in the case of systems governed by the heat-exchange principle, the voluntary action acts on the time constant of the governing first-order system of the process. Such "parametric" or "variable structure" systems have been studied extensively to analyse the stability problems encountered in adaptive control. However, they may sometimes exhibit unusual behaviour.

Consider a second-order system whose input $u(t)$ is a constant and whose output is $y(t)$. The natural frequency $\omega(t)$ varies with time or is dependent on a system variable that itself is a function of time.

$$\frac{\mathrm{d}^2 y(t)}{\mathrm{d}t^2} + 2\zeta\omega(t)\frac{\mathrm{d}y(t)}{\mathrm{d}t} + \omega(t)^2 \, y(t) = u(t)$$

where ζ and $\omega(t)$ are always positive. As $\omega(t)$ is always positive, the poles $\omega(t_0)$ are instantaneously stable. But surprisingly, depending on the frequency spectrum of the variation of the position of $\omega(t_0)$, it may be shown that the overall system may be unstable, although the "frozen" poles $\omega(t=t_0)$ are stable.

Such processes, which should be approached with caution, are encountered in the critical case where $u(t)$, a controlled variable or disturbance, acts on the structure and input simultaneously, as is the case with some self-tuning processes.

The simplest example of processes with real poles is that of a first-order system with a variable, positive, non-zero time constant $T(t)$. In this case, open-loop instability will not occur. Consider a process $y(t) > 0$ whose output, with zero input, is given by:

$$T(t)\frac{dy(t)}{dt} + y(t) = y_1 \text{ where } y(t=0) = y_0 > y_1 \; .$$

The behaviour of $D(t) = (y(t) - y_1)^2$, which represents the energy content, is given by:

$$\frac{dD(t)}{dt} = 2y(t)\frac{dy(t)}{dt} = -2y(t)\left[\frac{y(t)-y_1}{T(t)}\right].$$

Thus, $dD(t)/dt$ is always finite and negative as $y(t)$ goes from y_0 to y_1. Consequently, the controlled process is always stable for any positive variation of the time constant $T(t)$. In this case, there is no risk of "parametric instability" of the controlled process.

8.2 Heat Exchanger

Many industrial processes are based on the heat exchanger whose principle of operation may be approximated as shown in Figure 8.1. The diagram illustrates a heat-transfer fluid flowing through a double jacket with an input temperature T_e, volumetric flow rate F, density ρ_s, specific heat capacity Cp_s and volume V_s. The fluid is required to heat or cool the vessel contents, of density ρ_m and specific heat capacity Cp_m, to a temperature T_m. The vessel volume V_m exchanges heat with the fluid through a surface A with a heat-transfer coefficient U, a function of flow[10] (see Appendix B), and T_s is the output temperature of the heat-transfer fluid (see Figure 8.1).

---

[10]   Proposed by Benjamin Schramm of EVONIK.DEGUSSA.

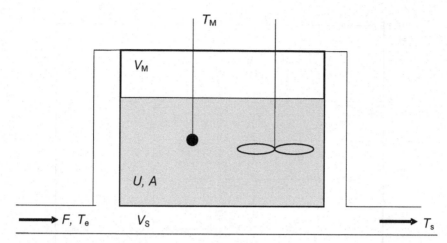

Figure 8.1. Heat exchange between a jacket and a vessel

A classical academic approximation of the heat-balance equation may be established by equating the heat-transfer fluid temperature to the output temperature of the double jacket:

$$\rho_m Cp_m V_m \frac{dT_m}{dt} = UA\left(T_s - T_m\right), \text{ and}$$

$$\rho_s Cp_s V_s \frac{dT_s}{dt} = UA\left(T_m - T_s\right) + \rho_s Cp_s F\left(T_e - T_s\right).$$

This approximation is valid only when the heat-transfer fluid flow rate is high. A more realistic analysis will be presented in Chapter 10. In most cases, the volume V_m is larger than V_s, resulting in one real pole dominating the second-order process[11]. In these circumstances the governing system equation may be approximated using a first-order model. Eliminating T_s from these equations, it may be shown that the system can be represented by a dominant pole resulting in:

$$\tau(F)\frac{dT_m}{dt} + T_m = T_e,$$

where $\tau(F) = \dfrac{a+bF}{F}$, $a = \dfrac{\rho_m Cp_m}{\rho_s Cp_s} V_m + V_s$ and $b = \dfrac{\rho_m Cp_m V_m}{UA}$. Thus, the flow rate acts on the process time constant (see Figure 8.2).

---

[11] It is assumed that the poles of any passive heat-exchange-based physical process are real ("the real poles heat-transfer conjecture").

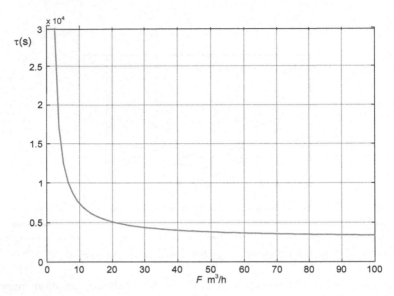

Figure 8.2. Time constant $\tau(F)$ as a function of flow rate F

The physical interpretation is clear. Consider the following example. The fluid in the vessel has a temperature of 30 °C and a hot/cold fluid temperature of 5 °C. Various constant flow rates $F(2, 10, 80 \ \text{m}^3/\text{h})$ are applied, resulting in the subsequent behaviours for T_m observed in Figure 8.3.

If the flow rate is zero, the vessel fluid temperature T_m does not vary and consequently the time constant is infinite. The resulting time constant would be large if the flow rate is small as there would be little exchange of heat. As the flow rate increases toward infinity the temperatures T_e and T_s become equal. In this case, the system equation may be described as:

$$\frac{\rho_m C p_m V_m}{UA} \frac{dT_m}{dt} + T_m = T_e = T_s \ \text{ as the time constant } \tau(F) \to b \text{ as } F \to \infty.$$

Assume that the reactor vessel contains water with a volume $V_m = 20 \ \text{m}^3$. An exothermic chemical reaction takes place and the product is cooled by a hot/cold fluid (water). The water has a temperature of 5 °C and a variable flow rate F with a volume $V_s = 0.7 \ \text{m}^3$. The fluid exchanges heat through a surface $S = 45 \ \text{m}^2$ with a heat-transfer coefficient $U = 1,\ 700 \ \text{W/m}^2\text{K}$. The typical characteristics for water at 20 °C are also assumed as:

$$\rho = 998.2 \ \text{kg m}^{-3} \text{ and } C_p = 4182 \ \text{J / kg / K} \ .$$

It may be shown that $a = 20.7$ m^3 and

$$b = (998.2)(4\ 182)(20/700)(45) = 2650 \text{ s} .$$

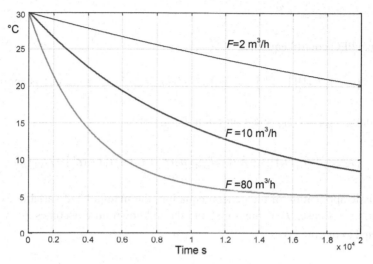

Figure 8.3. Temperature profile/flow

Consequently, $\tau(F) = (20.7 + 2650F)/F$, where F is expressed in m$^3/h$. Figure 8.3 shows that a flow rate $F = 2.5$ m^3/h results in a time constant $\tau \cong 30$ s and $F = 32$ m^3/h results in $\tau = 5$ s. Note that large changes in the system time constant arise for small flow rates. In the case where the exothermic reaction rate is effectively constant, a PI controller is sufficient, because only a small nominal change in the hot/cold fluid flow rate extracting the energy is required under these conditions. Thus, the dynamics vary only a little.

In the chemical industry, it is common to regulate the temperature T_e only when the quality requirement is not an issue. Such an approach results in an "average" regulation of T_m. Conversely, if the exothermic reaction changes rapidly, as is the case with polymerisation (energetic at the beginning and reducing towards zero at the end of the reaction), the subsequent variations of flow F will span all possible values.

A pertinent question that may be posed at this point is: *"How do you tune a PID regulator that has to control a process whose dynamics may vary by a factor of 20?"* The usual solution is to select a *"sluggish"* tuning scheme. While this procedure may not be satisfactory, it demonstrates no "obvious" evidence of T_m becoming unstable.

A first approach to solving such a problem would be to regulate the temperature of T_e while maintaining a constant flow F. This is a classic example of a cascade control strategy, as described in Chapter 7, being applied to regulate T_e and T_m .

A second approach involves acting on a flow F for a given temperature T_e. F must be either 0 or positive, because a negative (or cold) flow does not heat. The regulator will only function correctly for the condition:

$$T_m \geq T_{\text{Setpoint}} > T_e \ .$$

The usual PFC control equation:

$$\left(C - Y_p(n)\right)\left(1 - \lambda^h\right) = Y_m(n)\alpha^h + U(n)K_m\left(1 - \alpha^h\right) - Y_m(n) \text{ is applied.}$$

Substituting $T_{\text{mod}} = T_{\text{model}}$ and a set-point T_{Setpoint} gives:

$$\left(T_{\text{Setpoint}} - T_m(n)\right)\left(1 - \lambda^h\right) = T_{\text{mod}}(n)\alpha^h + T_e(n)1\left(1 - \alpha^h\right) - T_{\text{mod}}(n) \ .$$

In a classic application of the general equation, the manipulated variable would be T_e. But, in this case, T_e is measured and the unknown now becomes α, which is connected to the flow by:

$$\alpha = e^{-\frac{T_s}{\tau(F(n))}}, \text{ where } \tau\left(F(n)\right) = \frac{a + bF(n)}{F(n)} \ .$$

The control equation may be stated as:

$$\left(T_{\text{Setpoint}} - T_m(n)\right)(1 - \lambda^h) - T_e(n) + T_{\text{mod}}(n) = \alpha^h\left(T_{\text{mod}}(n) - T_e(n)\right) \ .$$

Extracting α^h gives:

$$\alpha^h = 1 - \frac{\left(T_m(n) - T_{\text{Setpoint}}\right)}{T_{\text{mod}}(n) - T_e(n)}\left(1 - \lambda^h\right) = 1 - A \ .$$

α must be less than or equal to 1 in order to ensure the flow is achievable. T_{mod} must not equal $T_e(n)$. It is not possible to use an "independent model" approach (see Chapter 2) as $T_{\text{mod}}(n)$ may vary significantly with respect to $T_m(n)$. Thus, the only remaining alternative is to use a realigned model, i.e., $T_{\text{mod}}(n) = T_m(n)$. In this case, we have:

$$\alpha^h = 1 - \frac{\left(T_m(n) - T_{\text{Setpoint}}\right)}{T_m(n) - T_e(n)}\left(1 - \lambda^h\right) \ .$$

We note that $T_m(n) - T_e(n) > T_m(n) - T_{\text{Setpoint}}$ and $(1 - \lambda^h) < 1$. This results in a solution for α^h, which is defined as α_{sol}. From this equation we get α_{sol} :

$$\tau_{\text{sol}} = -\frac{T_s}{\text{Log}(\alpha_{\text{sol}})} \text{ and } F_{\text{sol}} = \frac{a}{\tau_{\text{sol}} - b},$$

given that,

$$F = 0, \ \tau = \infty, \ \alpha = 1 \text{ and}$$

$$F = F_{\text{max}}, \ \tau_{\text{max}} = \frac{a + bF_{\text{max}}}{F_{\text{max}}}, \ \alpha_{\text{max}} = e^{-\frac{T_s}{\tau_{\text{max}}}},$$

then α has a value between 1 and α_{max} for $F = 0$ and $F = F_{\text{max}}$, respectively.

It is worth noting that the use of a realigned model will result in a biasing problem (see Chapter 2), which will be dealt with subsequently. Note that great care must be taken to respect the temperature hierarchy $T_m \geq T_{\text{Setpoint}} > T_e$.

The measurements may be corrupted by noise. Consequently, it is necessary to analyse the measurements using a logical filter whose goal is to respect the hierarchy. Otherwise, the control program will halt, presenting a solution that is mathematically not feasible, *i.e.*, $\alpha > 1$ or $\alpha < 0$.

Proceeding in this fashion results in a non-linear controller regulating a non-linear process by acting on a parameter of the process. Thus, parametric predictive control (PPC), as the name implies, acts on a process parameter, while PFC acts on a state variable of the process.

8.3 Constraint Transfer in Parametric Control

It is necessary to transfer the acting constraint F to the time constant T_{Setpoint} such that it is compatible with the physical system. From the control equation, $T_{\text{cons max}}$ may be determined:

$$\frac{(T_m(n) - T_{\text{Setpoint}})}{T_m(n) - T_e(n)} (1 - \lambda^h) = 1 - \alpha_{\text{max}}^h \ .$$

Substituting, $T_{\text{cons max}}$ for T_{Setpoint} gives:

$$T_m(n) - T_{\text{cons max}} = \frac{1 - \alpha_{\text{max}}^h}{1 - \lambda^h} (T_m(n) - T_e(n)),$$

and solving for $T_{\text{cons max}}$ results in:

$$T_{\text{cons max}} = T_m(n) - \frac{1 - \alpha_{\text{max}}^h}{1 - \lambda^h} (T_m(n) - T_e(n)) \ .$$

For $h=1$, the process loop will behave as a requirement on the reference trajectory with $\lambda = e^{-\frac{3T_s h}{\text{TRBF}}}$. The transfer function between T_{Setpoint} and T_m is given by:

$$\frac{T_m(s)}{T_{\text{Setpoint}}(s)} = Q(s) = \frac{1}{1+s\dfrac{\text{TRBF}}{3}}, \text{ where } Q(s) \text{ is the internal model.}$$

To prevent any bias arising from the use of a realigned model a transparent controller is introduced (see Figure 8.4) and F_{max} is transferred to the MV as the constraint $T_{\text{cons max}}$.

Figure 8.4 represents a vessel containing a mass of fluid whose temperature T_m is initially 50 °C. The objective is to maintain the temperature of the vessel at a set-point of 20 °C. At first, the hot/cold fluid flow (whose temperature is 5 °C) in the jacket reaches the maximum flow rate of 30 m³/h and subsequently reduces towards zero as the mass temperature approaches the set-point. (It is assumed that there is no heat loss or gain in the reactor.)

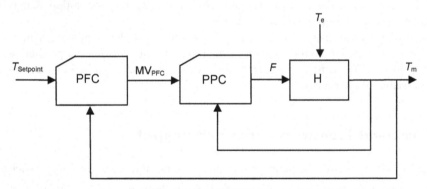

Figure 8.4. PPC/PFC flow control

Figure 8.5 shows results from the parametric control of the vessel. At 300 s, a disturbance caused by an exothermic reaction occurs and the vessel fluid temperature increases, resulting in a fluid flow rate that stabilises the system at 10 m³/h. At 600 s, another disturbance of a similar nature results in a hot/cold fluid flow rate at the maximum limit of 30 m³/h at steady state. It may be observed that the output MV_{PFC} of the outer PFC regulator presents an "intermediate" set-point to the PPC controller. At the end of the testing phase, this set-point drifts as the regulating PPC reaches its maximum available set-point flow (flow constraint).

8.4 Evaluation

Flow control, using a non-linear regulator, is quite advantageous if a source of heat-transfer fluid is available, because only *one valve* will suffice, eliminating the necessity for any approximation of the non-linear process. It is recommended that a flow controller is used to deal with the usual valve defects.

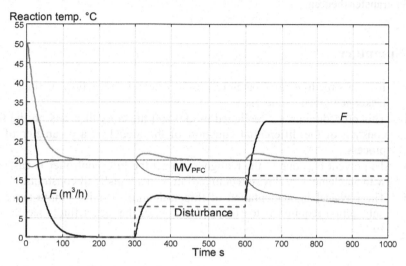

Figure 8.5. Parametric control of a reactor. PPC/PFC temperature control using fluid flow rate

However, this method has its limits. For instance, *how does the process behave in the event of a disturbance occurring in the reactor?* If the flow is large when the disturbance occurs, the response is rapid. Conversely, if a small flow exists at the time of disturbance, the dynamics of the initial response will also be slow and the controller will have a sluggish start. Thus, it is advisable to always maintain some basic dynamics, *i.e.*, preserve a non-zero minimum flow F_{min} .

If we again consider the enthalpy balance describing the hot/cold fluid equation it may be seen that it consists of the product $F(T_e - T_s)$ (see Figure 8.6). If F is increased to achieve a fast process response then the term $T_e - T_s$ must be decreased. Thus, in the cooling case, T_e is required to converge on T_s. This may be achieved by feeding back a portion of the jacket fluid outflow, whose temperature is $T_s > T_e$, to the input. The advantage of this approach is that no other heat source is required.

So, we may act on the flow using PPC controller $(T_e = \text{constant})$ or we may act on the temperature using a PFC controller $(F = \text{constant})$. Alternatively, we may regulate the thermal power by simultaneously manipulating the two components F and $(T_e - T_s)$.

However, F and T_e are linked because they come from a source that, in general, has non-zero internal impedance. Other considerations must be taken into account if *enthalpy control*, with the two components, is to be used. The two internal impedances of the source and the reactor, which are linked differently to F and T, should be modified. This is another example of the classical problem of impedance matching similar to that seen in electrical networks, *i.e.*, the maximum power-transfer theorem.

8.5 Summary

- It is possible to control batch reactors by acting on the flow of fluid in the jacket at a given temperature.
- This procedure uses a realigned model and the controller acts on the time constant of the differential equation of the model – a parameter of the process.
- To prevent bias due to the realigned model the PPC controller is cascaded with an elementary PFC in combination with a transfer of constraint(s).
- The physical implementation of this technique is quite simple since the only action required is to control the flow of the jacket fluid (see Section 10.5).

9

Unstable Poles and Zeros

Abstract. This chapter deals specifically with the problem of processes containing unstable poles and inverse response zeros. The relative positions of the poles and zeros have a direct impact on the level of the stability of the system and thus dictate the complexity of the controller. Fortunately, these cases are rare. However, such systems are becoming more commonplace as modern integrated design techniques use physical systems that are manufactured intentionally unstable in order to maximize the performance envelope of the controlled process.

This chapter may be omitted if this book is being read to develop an understanding of the predictive control concept only.

Keywords: Inverse response, non-minimum phase, robustness, stability, tuning

9.1 Complexity

In Chapters 4 and 8 it was shown that controlling an unstable system with a direct response (*i.e.*, no inverse response) did not present much difficulty when the decomposition principle was applied. The same may be said for stable processes having a negative zero, referred to as a *non-minimum phase*, *inverse response* or *unstable zero* process. The only caveat is that the coincidence point must be placed after the inverse response "dip" if the process is to be controlled successfully.

However, processes that are both unstable and non-minimal phase are more difficult to stabilise [18]. The study of such systems is not just an academic exercise since such systems are encountered in practice.

For instance, consider the case of a chemical reactor where the introduction of a cold reagent induces a physical cooling effect. This effect begins before the onset of an exothermic chemical reaction that, if left unchecked, would drive the system unstable.

This situation is also often encountered in lightly damped mechanical systems. Due to inertia, the control of such systems results in a behaviour where the initial motion is contrary to the final motion. Generally, in aeronautics and for planes and missiles in particular, the transfer functions typically possess poles and unstable zeros. The general transfer function $H(s) = \dfrac{1+sT_1}{1+sT_2}$ will be considered here, where $T_1 < 0$ and $T_2 < 0$. To reinforce this, the problem is standardised by setting $T_2 = -100$ s and allowing T_1 to vary.

Under these circumstances, the concepts of steady state and transient response lose their usual meaning. The equivalent steady state gain is assumed strictly unitary, *i.e.*, an $MV = 1$ in steady state would result in $CV = 1$, which does not exist in practice. The effective response is unstable and depends mainly on the values of T_1 relative to T_2. For a step input $MV(n) = 100$ the output is:

- positive for $T_1 = -110 < T_2 < 0$;
- negative for $T_2 < T_1 = -50$;
- unitary for $T_1 = T_2$.

Figure 9.1 illustrates this behaviour, with $T_1 = -110$ s and $T_1 = -50$ s .

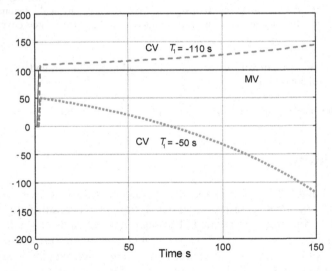

Figure 9.1. Characteristic open-loop response $H(s) = \dfrac{1+sT_1}{1-s100}$

The transfer function $H(s)$ can be decomposed by setting $A = (T_1 / T_2)$ and $B = 1 - A$, giving: $H(s) = A + \dfrac{B}{1+sT_2}$.

In discrete form: $H(z^{-1}) = Az^{-1} + \dfrac{B(1-a_2)z^{-1}}{1-a_2 z^{-1}}$ in discrete form. We will proceed by analysing the example considered in previous chapters.

9.2 Stable Pole and Stable Zero

A stable pole and stable zero in $H(s) = \dfrac{1+sT_1}{1+sT_2}$ results in positive values for T_1 and T_2. The output of the decomposed forward-path transfer function of gain A (see Section 9.1) is equal to the sum of the zero free response and the forced response output to a step input of amplitude A. The combined forward-path transfer-function output is given by:

$$s_{m1}(n) = MV(n-1)A .$$

The dynamic transfer function output is given by:

$$s_{m2}(n) = s_{m2}(n-1)a_2 + B(1-a_2)MV(n-1) .$$

The sums of the free and forced outputs are:

$$S_L = 0 + s_{m2}(n)a_2^h \text{ and } S_F = (A + Bb_{2h})MV(n-1) ,$$

where $b_{2h} = 1 - a_2^h$.

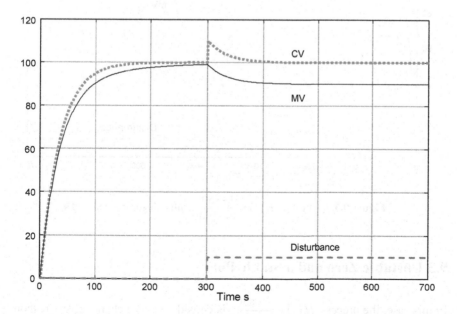

Figure 9.2. Control of $H(s) = \dfrac{1+s110}{1+s100}$

Given that h is the coincidence point, $l_h = 1 - e^{-\frac{3T_s h}{CLTR}}$ the following may be stated:

$$MV(n) = \frac{(Setpoint - CV(n))l_h + s_{m1}(n) + s_{m2}(n)b_{2h}}{A + Bb_{2h}}.$$

Assuming that $T_1 = 110$ s, $T_2 = 100$ s, $h = 1$, $A = 3$, $B = 1$ and CLTR = 100 s, the resulting behaviour is shown in Figure 9.2. A strong overshoot results in open loop for a forward-path gain of 0.3. Thus, it is important to choose $h = 1$. Also note that the MV has a sluggish start.

The behaviour for $T_1 = 50$ s is shown in Figure 9.3 using the same tuning parameters. For different values of T_1, it may be seen that the rate of change of the response is very diverse for a change in set-point.

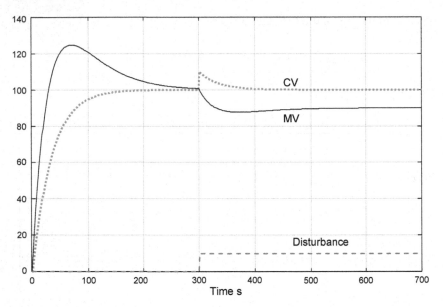

Figure 9.3. Control of $H(s) = \dfrac{1+s50}{1+s100}$ with a disturbance at 300 s

9.3 Unstable Zero and a Stable Pole

In this case, the process $H(s) = \dfrac{1-s50}{1+s100}$ is considered. The characteristic response of the system is shown in Figure 9.4 where the "dip" of the inverse response persists for a duration of 45 s. A satisfactory response is achieved without any particular difficulty when a coincidence point beyond this stage is used.

For $h = 50$ and $CLTR = 10\,s$, the closed-loop response is similar to that of the open loop where h is large. But the resulting effective CLTR is not as specified (see Figure 9.4).

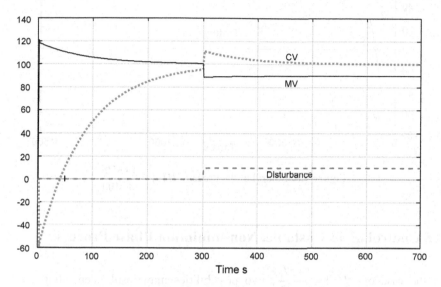

Figure 9.4. Control of the process $H(s) = \dfrac{1-s50}{1+s100}$

9.4 Control of an Unstable, Minimum Phase Process

As the process $H(s) = \dfrac{1+s50}{1-s100}$ is unstable, it may be decomposed in the usual way. Using the notation of Chapter 2, the model for the forward path is given by:

$$M_1(s) = -2\frac{1+s50}{1+s200}, \text{ and the feedback path by: } M_2(s) = \frac{K}{1+s200},$$

where K is calculated using $1+s200 - K = -2(1-s100)$ with $K = -3$ (see Chapter 2).

Assuming $h = 55$ and $CLTR = 100\,s$, Figure 9.5 shows an MV proceeding initially in a direction that is opposite to its final value. The disturbance that occurs at 500 s is rejected. Also note that the internal model of the regulator is very stable.

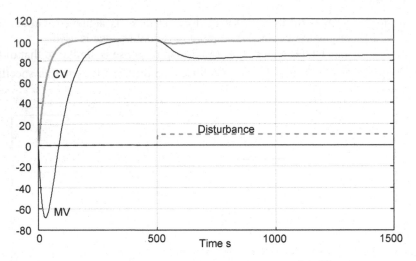

Figure 9.5. Control of the process $H(s) = \dfrac{1+s50}{1-s100}$

9.5 Control of an Unstable, Non-minimum Phase Process

In the case of $H(s) = \dfrac{1+sT_1}{1+sT_2}$, two possibilities may result, depending on the relative values of T_1 and T_2 where both are assumed to be negative. Only the case where $T_2 < T_1 < 0$ is considered here.

For reasons of implementation and initialisation, it is necessary that the closed loop and the regulator be stable. This implies that the internal models of the regulator must also be stable. Disturbances must be rejected and a certain level of robustness must be maintained in the presence of structural mismodelling. All existing elementary techniques that ignore the effects of the unstable poles of the model, as a result of cancellation, are not acceptable. The chosen procedure should employ a decomposition technique that uses only stable models.

The discrete simulation of a transfer function of order 1/1 poses problems. A filter of time constant τ may be added to simplify the simulation. This time constant can be interpreted as taking the dynamics of the actuator into account. It may be either physically added to the input or included as a process calculation.

The values CLTR $= 0$ and $h = \infty$ are adopted to reduce the number of degrees of freedom and to simplify the tuning of the regulator. Thus, the remaining tuning parameters are the time constants of the decomposition τ and θ (see Chapter 2). Stability may be calculated in the continuous domain where τ and θ are positive.

The forward-path transfer function M_0 (see Figure 2.2) as shown in Figure 9.6 is given by $M_0(s) = \dfrac{1+sT_1}{(1+sT_2)(1+s\tau)}$. The decomposition procedure of Chapter 2 may be applied resulting in a forward-path transfer function of $M_1(s) = (1-K)\dfrac{1+sT_1}{(1+s\theta)(1+s\tau)}$ and a feedback-path transfer function of

$M_2(s) = \dfrac{K}{1+s\theta}$. If $K = 1 - \dfrac{\theta}{T_2}$ then it may be shown that $M_0(s) = \dfrac{M_1(s)}{1 - M_2(s)}$.

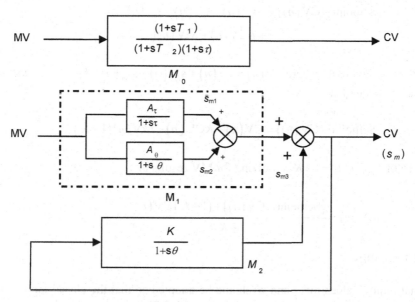

Figure 9.6. Decomposition of the model $H(s) = \dfrac{1+sT_1}{(1+sT_2)(1+s\tau)}$

The output of the forward-path transfer function M_1 is denoted by $s_{m1} + s_{m2}$ while the output of the feedback-path transfer function M_2 is denoted by s_{m3}. The decomposition of the forward-path transfer function into more fundamental elements is given by:

$$s_m(n) = s_{m1}(n) + s_{m2}(n) + s_{m3}(n) .$$

In the nominal case $s_m(n) = s_p(n)$,

$$s_{m2}(n) = s_{m2}(n-1)a_\tau + b_\tau A_\tau \text{MV}(n-1), \quad \text{and}$$

$$s_{m3}(n) = s_{m3}(n-1)a_\theta + b_\theta A_\theta \text{CV}(n-1),$$

where $A_\theta = \dfrac{T_1 - \tau}{\theta - \tau}$, $A_\tau = 1 - K - A_\theta$, $a_\theta = e^{-\frac{T_s}{\theta}}$, $b_\theta = 1 - a_\theta$, $a_\tau = e^{-\frac{T_s}{\tau}}$, $b_\tau = 1 - a_\tau$.

If $h \to \infty$, the free output responses of type $S_1 = a^h$ become zero (since $a < 0$). The forced output responses become $MV(n)A_\theta$ and $MV(n)A_\tau$ for the forward-path transfer function and $CV(n)K$ for the return-path transfer function.

Similarly, $l_h = 1 - \lambda^h \to 1$. The control equation becomes:

$$\left(\text{Setpoint} - CV(n)\right) = -s_{m1}(n) - s_{m2}(n) - s_{m3}(n)$$
$$+ KCV(n) + MV(n)\left(A_\theta + A_\tau\right) \; .$$

In the nominal case, where $CV(n) = s_{m1}(n) + s_{m2}(n) + s_{m3}(n)$, the control equation may be re-stated as:

$$\left(\text{Setpoint} - CV(n)\right) = CV(n) + KCV(n) + MV(n)(1 - K) ,$$

given that $A_\theta + A_\tau = 1 - K$. Solving for MV results in:

$$MV(n) = \frac{\left(\text{Setpoint} - CV(n)\right) + (1 - K)CV(n)}{1 - K} \; .$$

9.5.1 Stability

What tuning values for τ and θ should be adopted so that the closed-loop system poles are real and the resulting system is stable? The stability calculation may be determined in the continuous domain by noting that:

$$CV(s) = \frac{(1 + sT_1)}{(1 + sT_2)(1 + s\tau)} MV(s), \text{ and } MV(s) = \frac{(1 + sT_2)(1 + s\tau)}{1 + sT_1} CV(s) \; .$$

Substituting into the control equation gives:

$$\text{Setpoint} - CV(s) = -(1 - K)CV(s) + (1 - K)\frac{(1 + sT_2)(1 + s\tau)}{1 + sT_1} CV(s).$$

Finally, the closed-loop transfer function is determined by replacing K with $1 - \dfrac{\theta}{T_2}$ giving:

$$\frac{CV(s)}{\text{Setpoint}} = \frac{1 + sT_1}{1 + s\left[T_1 + \dfrac{\theta}{T_2}\left(-T_1 + T_2 + \tau\right)\right] + s^2 \tau \theta} \; .$$

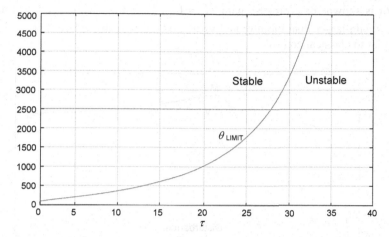

Figure 9.7. Stability of the control of the process $H(s) = \dfrac{1 - s50}{(1 - s100)(1 + s15)}$

The poles will be real and stable if:

$$\left[T_1 + \frac{\theta}{T_2}(-T_1 + T_2 + \tau) \right]^2 - 4\tau\theta > 0,$$

where τ is a positive variable that must be chosen. The previous expression is developed and the limiting value for θ, *i.e.*, the stability bounds, is determined. The solution is given by the following equation:

$$\frac{(-T_1 + T_2 + \tau)^2}{T_2^2}\theta^2 + 2\left[\frac{T_1}{T_2}(-T_1 + T_2 + \tau) - 2\tau \right]\theta + T_1^2 = 0 .$$

For each value of τ such that $0 < \tau < T_2 - T_1$ the bound may be determined using the equation $\theta = \theta\lim(\tau)$ which ensures that the poles are real (see Figure 9.7). The stability domain, expressed in the $\theta - \tau$ plane, is shown for $T_1 = -50$ s and $T_2 = -100$ s. For $\tau = 15$ s, the bound is given by $\theta\lim(15) = 748.23$ s. Taking $\theta = 550$ s results in a system with real poles.

9.5.2 Robustness

Figure 9.8 shows the ability of the system to absorb the effects of an additive disturbance at 1000 s. Figure 9.9 illustrates the behaviour of this process for a variable process gain set to a value of $G_p = 0.90$, as opposed to the internal model gain $G_m = 1$ with $\tau = 5$ s and $\theta = 550$ s, while in Figure 9.10 the response of the system is shown for a process gain $G_m = 1.1$. Note that, within certain limits, a process gain larger than the nominal value has a stabilising effect. This behaviour is similar to classical control.

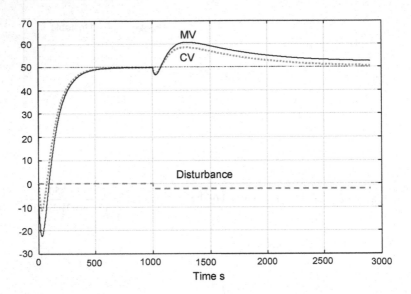

Figure 9.8. Control of the process $H(s) = \dfrac{1-s50}{(1-s100)(1+s15)} G_p$, where $G_p = G_m = 1$

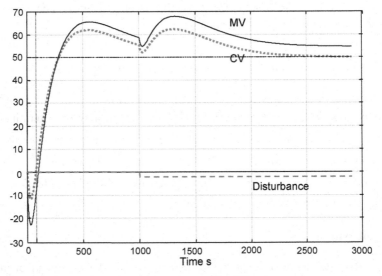

Figure 9.9. Control of the process $H(s) = \dfrac{1-s50}{(1-s100)(1+s15)} G_p$, where $G_p = 0.90$ and $G_m = 1$

9.5.3 Tuning

The values $\tau = 5$ s and $\theta = 500$ s are selected using a mathematical analysis technique where stability is the primary objective. Thus, the dynamic performance, *i.e.*, the CLTR, is not directly controlled as is the case in the previous examples. It is simply a consequence of tuning. The mathematical, closed-loop system model can be easily determined with the help of the equation:

$$\frac{CV(s)}{Setpoint} = \frac{1+sT_1}{1+s\left[T_1 + \dfrac{\theta}{T_2}\left(-T_1 + T_2 + \tau\right)\right] + s^2\tau\theta}.$$

Two strategies may be implemented:

1. The pair τ, θ may be fixed. This eliminates the need to simulate the process and its regulator loop. It is necessary only to simulate the process response as defined by the above equation. A change in set-point is then applied and the resulting CLTR is noted. Figure 9.11 shows the behaviour of the CLTR for different values of τ and θ. The resulting linear behaviour is caused by the dominant pole.
2. Alternatively, the CLTR may be fixed for a given value of θ (*i.e.*, a large value) and an acceptable value of τ is sought.

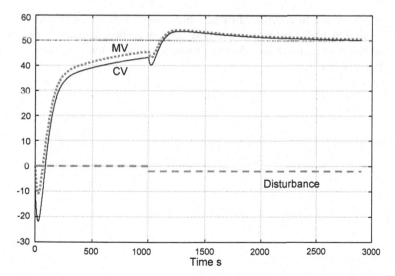

Figure 9.10. Control of the process $H(s) = \dfrac{1 - s50}{(1 - s100)(1 + s15)} G_p$ with $G_p = 1.1$ and $G_m = 1$

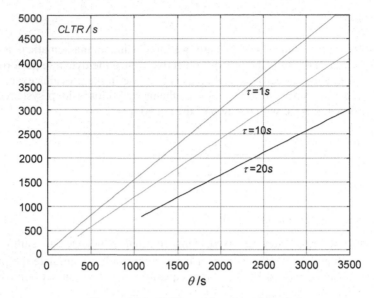

Figure 9.11. CLTR as a function of τ and θ

9.6 Summary

- It is possible to control "difficult" processes, with unstable and with non-minimum phase behaviour (inverse response), provided the time constant of the numerator is less than the time constant of the denominator of the transfer function.

10

Industrial Examples

Abstract. Since its introduction in 1974, the number of successful applications of the PFC technique has grown into the thousands. In this chapter some of these applications are presented to illustrate the use of the techniques discussed in the previous chapters of this book. The examples represent typical problems encountered in the process industry and deal with industrial units that may be described as interactive SISO processes. Heat exchange is a basic issue encountered in many industries, so the modelling of a heat exchanger is firstly addressed. Examples of controlling such systems are then discussed. A complete case study of level control at the Arcelormittal plant in Fos-sur-Mer is presented and deals with the well-known and difficult problem of continuous casting. The problem of steam generation at the Fos-Sur-Mer plant is then addressed. The next example illustrates the PPC control of a batch reactor at the EVONIK.DEGUSSA plant. A comparison with classical PID control is given that clearly demonstrates the technical and economic benefits associated with the introduction of the PFC technique.

Keywords: Heat exchange, counter-current exchanger, convexity, modelling, identification, superheater, desuperheater, level control, zone control, parametric predicitve control

10.1 Industrial Applications

Since 1974, many industrial applications of predictive control have been developed. Its use has spanned a diverse range of application areas, from fast mechatronics systems, such as robotics, to slow processes in the production industry (chemical, energy generation, *etc.*).

Thus, given such a broad application range, it is not surprising that difficulties arise when dealing with the practical issues surrounding implementation. Some of the issues commonly encountered include complex dynamics, constraints, non-linearity and environmental factors, such as disturbances and specific operating requirements. Also, the sampling times associated with such systems can vary from tens of microseconds (*e.g.*, weapon systems) to one hour (*e.g.*, maintaining river water level).

The discussion of practical case studies provides insight into the implementation of the PFC algorithm and helps highlight the potential of the techniques presented in the book.

The first case study deals with a pilot plant situated at the Institut de Régulation d'Arles. An industrial heat exchanger was chosen to illustrate the PFC/PPC (see Chapter 8) technique. The location of the unit in a teaching facility permitted all the necessary measurements and comparisons to be carried out without being restricted by the pressures of production.

The second case study takes place at the ARCELORMITTAL (formerly SOLLAC) steel plant in Fos-sur-Mer. This example highlights some of the issues that must be faced if a controller installation is to be successful on a long-term basis. Predictive control has been applied to several of the production units at this plant since 1986. Factors such as convenience of tuning and adapting to changing economic circumstances have proved crucial to the long-standing use of predictive controllers at the plant.

10.2 Heat Exchanger

The problem of modifying a product temperature under the action of a heat-transfer fluid is commonly encountered in industry. Many established industrial techniques exist for dealing with such situations. In the following sections we discuss one such approach to solving this problem [3, 6, 13].

10.2.1 Problem Description

Heat transfer is a fundamental requirement of most operations in the process industry. Consequently, the heat exchanger, in its various configurations, is commonplace. There are numerous comprehensive texts available on this subject, primarily concerned with explaining the heat-transfer concept and the physical design of such devices. Historically, little attention has been paid to dynamic modelling for the purpose of control.

In most cases, the behaviour of the process does not pose any major problems. Thus, a "loosely" tuned PID controller is generally adequate. However, as requirements increase, industry demands more effective controllers that are easily tuned. Also, the ability to compensate for variations in flow rate, product and heat-transfer fluid input temperatures is considered particularly important.

Here, the common case of the counter-current exchanger is considered. The objective is to regulate the temperature of the output fluid, the product, by acting on the counter-current flow rate of the "heat-transfer fluid" (see Figure 10.1).

The product enters the exchanger at a flow rate Q_p and temperature t_{ep} and leaves with a temperature t_{sp}. At the opposite end of the device, the heat-transfer fluid enters with a flow rate Q_f and temperature t_{ef} and exits with a temperature t_{sf}.

The goal is to ensure that tsp is maintained at a fixed set-point by acting on the manipulated variable Q_f. It is possible that Q_p, t_{ep} and t_{ef} (i.e., all the process variables) may vary at the same time. In such circumstances, the resulting process is variable and non-linear with a pure time delay, whereas the temperature t_{sp} depends on the variables Q_p, t_{ep} and t_{ef}. When Q_p, t_{ep} and t_{ef} vary with time, using a PID regulator with fixed parameters would result in inefficient operation or, at worst, instability. Also, if the exchanger is physically large, the tuning time may become unacceptably long. In such cases, a controller using a knowledge-based (first-principles) model, that is easy to tune, has definite advantages.

Before modelling this process, it is advisable to consider some less complex heat-exchanger examples.

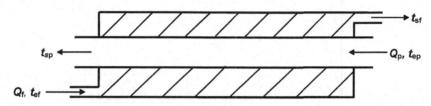

Figure 10.1. Counter-current exchanger

10.2.2 Convexity Theorem

The convexity property is a simple and convenient way to represent the phenomenon of heat exchange for the purposes of dynamic modelling and control. Consider the simple case where two similar fluids are mixed, resulting in a combined temperature T and heat flow rate F (see Figure 10.2). Assume that the process is lossless and possesses no mechanical work influence or internal friction.

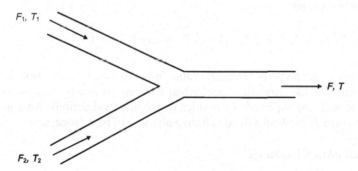

Figure 10.2. Mixture of two identical fluids F_1, T_1 and F_2, T_2

Assuming the fluid heat capacity remains constant over a temperature range $T_1 \leftrightarrow T_2$, the resulting mass and energy balance equations may be written as:

$$FT = F_1T_1 + F_2T_2 \ .$$

Extracting T gives $T = T_1 \dfrac{F_1}{F} + T_2 \dfrac{F_2}{F}$, and defining $\lambda = \dfrac{F_1}{F}$ gives

$T = \lambda T_1 + (1-\lambda)T_2$. The coefficient λ, referred to as the convexity coefficient, is assumed to have a value between 0 and 1 (see Figure 10.3).

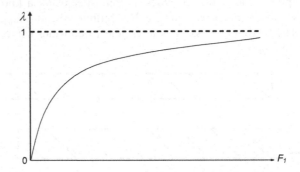

Figure 10.3. Convexity coefficient as a function of flow rate F_1

In the equation describing T above, the temperature variations are linear, whereas λ depends on a flow rate that is governed by a hyperbolic function. The values of λ for the extreme cases, $\lambda=0$, $F_1 = 0$ and $\lambda = 1$, $F_2 = 0$ possess a physically intuitive interpretation.

The combined temperature T, between temperatures T_1 and T_2, possesses all the characteristics of mathematical convexity. Recall that a set **T** is considered convex if, for any two values:

$$\{T_1, T_2\} \in \mathbf{T},\ T = \lambda\, T_1 + (1-\lambda)T_2\ \text{ where } 0 \le \lambda \le 1\ .$$

This convexity property is normally valid only for steady state. But, it may be incorporated into the dynamic model when the various steady state measures are compatible with the expected convexities (instrument validation). An approach of this nature greatly facilitates the qualitative analysis of heat processes.

10.2.3 Fluid/Mass Exchange

Consider the slightly more complex, but nonetheless common, example of a fluid circulating in a double jacket where the temperature of the fluid contained within the vessel is considered uniform. The heat-transfer fluid in the reactor is treated as a single quantity on a consignment or "batch" basis. From a modelling perspective, this may be considered similar to treating a consignment (or a series of consignments) in a temperature-controlled treatment oven (see Figure 10.4). The temperature of the oven is assumed uniform throughout (see Figure 10.5). Also, the consignment, or fluid volume, is exposed to the heat of the oven for a finite period of time, *i.e.*, the time required to transport the quantity of fluid through the oven.

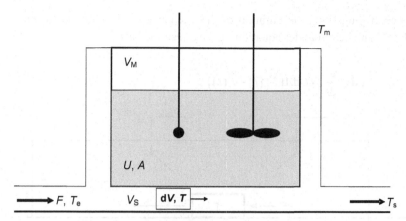

Figure 10.4. Batch reactor

The temperature of the heat-transfer fluid varies both temporally and spatially. Consequently, the heat-transfer phenomenon may be analysed either by (a) considering a small, elementary, mobile volume dV (see Figure 10.4) of heat-transfer fluid and developing the heat-balance equations for the various stages of transportation through the jacket, or (b) segmenting the heat-transfer fluid space into a large number of small, fixed volumes and developing the heat- and mass-balance equations for each of these elements.

The first approach is used for the heat treatment oven, while the second is used for the exchanger. The fluid possesses the following characteristics:

ρ Density $\left(\text{kg m}^{-3} \right)$

Cp   Specific heat capacity $\left(\text{J K}^{-1} \text{ kg}^{-1} \right)$

F Volumetric flow rate $\left(\text{m}^3 \text{ s}^{-1} \right)$

U Heat-transfer exchange coefficient $\left(\text{Wm}^2 \text{ K}^{-1} \right)$

A Heat-transfer area $\left(\text{m}^2 \right)$

V Heat-transfer fluid volume $\left(\text{m}^3 \right)$

A small fluid volume enters the heat exchanger at a temperature $T_{e0}(t)$ and exits at a time $t + \theta$ with a temperature $T_s(t + \theta)$, while the mass temperature $T_m(t)$ of the reactor (or oven) remains uniform along the heat exchanger. It is assumed that the exchange of heat is exclusively perpendicular to the direction of flux flow and that no lateral diffusion exists. The fluid section is also assumed constant and the volume $dV = V/N$ (where N is large) is sufficiently small such that it is always in contact with the exchange surface $\dfrac{dV}{dA} = \dfrac{V}{A}$ at a temperature $T_m(t)$. Assuming the presence of plug flow, the duration of the exchange θ depends on the flow rate $\theta = V/F$. The heat-transfer equation may be described as:

$$\rho \mathrm{Cpd}V\frac{\mathrm{d}T}{\mathrm{d}t} = U\mathrm{d}A\big(T_\mathrm{m}(t) - T(t)\big) .$$

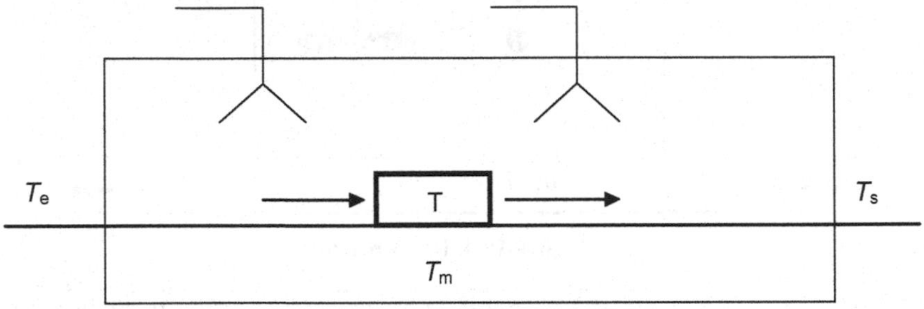

Figure 10.5. Heat-treatment oven

Replacing $\mathrm{d}V$ and $\mathrm{d}A$ with $\dfrac{V}{N}$ and $\dfrac{A}{N}$, respectively, provides the following simplification:

$$\rho \mathrm{Cp}V\frac{\mathrm{d}T}{\mathrm{d}t} = UA\big(T_\mathrm{m}(t) - T(t)\big), \text{ i.e., } N \text{ has been eliminated.}$$

The heat-transfer coefficient U varies with temperature and flow rate, as was observed in the counter-current exchanger case. The dependency of U on the magnitude of the flow rate may be taken into account. However, for simplicity, it is assumed that the coefficient is constant all along the jacket. Also, assume that the nature of the outflow does not vary, *i.e.*, is always turbulent. The process may be described by:

$$\tau \frac{\mathrm{d}T(t)}{\mathrm{d}t} + T(t) = T_\mathrm{m}(t), \text{ where } \tau = \frac{\rho \mathrm{Cp}V}{UA} .$$

Thus, T is the output of a first-order process subjected to an input $T_\mathrm{m}(t)$ applied for a finite period of time θ. Given that $\theta = \dfrac{V}{F}$ and $\tau = \dfrac{\rho \mathrm{Cp}V}{UA}$, we get:

$$\mathrm{e}^{-\frac{\theta}{\tau}} = \mathrm{e}^{-\frac{UA}{\rho \mathrm{Cp}F}} = \mathrm{e}^{-\mathrm{NUT}} .$$

If the flow rate F is assumed constant, the temperature $T(t)$ appears at the output of the exchanger with a value $T_s(t)$. This value depends on the initial condition $T_{e0}(t-\theta)$ and on the temperature $T_m(u)$, where $(t-\theta)<u<t$ during its passage through the exchanger.

The process may be analysed in discrete form. The small volume dV begins with an initial temperature T_{e0} and is subjected to temperature T_m during the discrete interval rT_{ech}, where T_{ech} represents the sampling time.

Defining $r = \text{round}\left(\dfrac{\theta}{T_{ech}}\right)$ and $a = e^{-\frac{T_{ech}}{\tau}}$ allows the process to be described in difference equation form as:

$$T(n) = T(n-1)a + (1-a)T_m(n-1) .$$

Iterating this equation from 1 to r gives:

$$n = 2: \ T(2) = T_{e0}a + (1-a)T_m(1)$$

$$n = r:$$

$$T(r) = T_{e0}a^{r-1} + (1-a)\left[T_m(r-1)a + T_m(r-2)a^2 + \ldots + T_m(1)a^{r-1}\right] .$$

It is possible to create a Laplace transform representation $T(s)$ of the solution $T(t)$. The solution $T(t)$ consists of a classical free term $T_1(t)$ and a forced term $T_f(t)$ (see Chapter 3). At the exchanger output we get:

Free term (T_1): $T_1(t) = T_{e0}(t-\theta))e^{-\theta/\tau} = \lambda T_{e0}(t-\theta)$ with a transportation lag θ and a decay $\lambda = e^{-\theta/\tau}$.

Forced term (T_f): It is convenient to temporarily use a discrete convolution form in order to better represent the physical manifestation of the forced term. The mass temperature $T_m(t)$ may be considered as acting on a unit gain filter with a time constant τ, but with a finite memory, corresponding to the duration of the transportation journey.

It is assumed that $\Theta = \left\lfloor \dfrac{\theta}{T_{ech}} \right\rfloor$ (with an integer number of periods of T_{ech}, where T_{ech} is the sampling time.

$$T_f(n) = 1.T_m(n) + aT_m(n-1) + \ldots + a^i T_m(n-i) + \ldots + a^\Theta T_m(n-\Theta),$$

where $a^i = e^{-\frac{T_{ech}\, i}{\tau}}$, $i = 1...\Theta$. A convolution representation, similar to that introduced in Section 4.9, is adopted, resulting in a weighting sequence of the form:

$$Z = \begin{bmatrix} 1 & a & a^2 ... a^i ... a^\Theta \end{bmatrix}, \text{with } a^\Theta = e^{-\frac{\theta}{\tau}}.$$

Given that $\theta = \dfrac{V}{F}$ and $\tau = \dfrac{\rho Cp V}{UA}$ results in: $e^{-\frac{\theta}{\tau}} = e^{-\frac{UA}{\rho CpF}} = e^{-NUT}$,

where NUT is usually denoted by the term "number of transferred units".

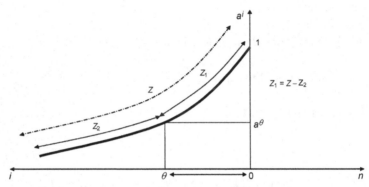

Figure 10.6. A finite convolution representation of 2 infinite sequences

In this case, the weighting sequence is finite since the output temperature does not depend on past values of the temperature T_m that follow a fixed path through the double jacket. Conversely, if a classical alternative is sought for the finite convolution representation Z_1 (see Section 4.9), it may be noticed that the finite weighting sequence Z_1 represents the difference between two infinite weighting sequences Z and Z_2:

$$Z_1 = Z - Z_2 \text{ (see Figure 10.6)}.$$

Z_2 corresponds to a system that represents a pure delay θ and a gain other than 1.

Thus, the forced output $T_{f1}(n)$ corresponds to the difference between the forced outputs $T_f(n)$ (sequence Z) and $T_{f2}(n)$ (sequence Z_2) given by:

$$T_{f2}(n) = T_{f2}(n-1)a + (1-a)a^\Theta T_m(n-1-\Theta) \text{ and } T_{f1}(n) = T_f(n) - T_{f2}(n).$$

The response represents the sum of three terms (one free and two forced response terms). Setting $\lambda = e^{-\frac{\theta}{\tau}} = a^{\Theta}$ results in the transfer-function representation shown in Figure 10.7.

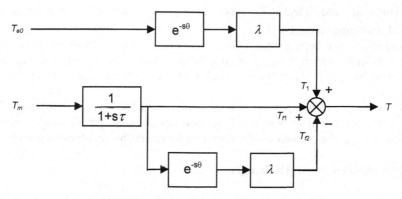

Figure 10.7. Transfer-function representation

Confining the analysis to the steady state results in a constant T_{e0}, T_m, and T_{f1}. $T_{f2} = T_{m0}e^{-\frac{\theta}{\tau}}$ and the required convexity is found:

$$T_s = T_{e0}\lambda + T_{m0}(1-\lambda), \text{ where } \lambda = e^{-NUT}.$$

As was seen in Chapter 8, the double-jacket zone containing the heat-transfer fluid may be considered as a single compartment V with a flow rate F and a temperature T_e. This results in:

$$\frac{\rho CpV}{UA + \rho CpF}\frac{dT}{dt} = \frac{UAT_m(t) + \rho CpFT_e(t)}{UA + \rho CpF}.$$

Defining $\lambda_{elem} = \dfrac{\rho CpF}{UA + \rho CpF}$ and $\tau_{elem} = \dfrac{\rho CpV}{UA + \rho CpF}$

gives $\tau_{elem}\dfrac{dT}{dt} + T = T_e(t)\lambda_{elem} + T_m(t)(1-\lambda_{elem})$.

Comparing terms gives:

$$\lambda = e^{-NUT} = e^{-\frac{\theta}{\tau}} = e^{-\frac{UA}{\rho CpF}} \text{ and } \lambda_{elem} = \frac{1}{1 + NUT}.$$

If the flow rate is large with respect to UA :

$$\lambda = 1 - \mathrm{NUT} + 0.5.\mathrm{NUT}^2 + ..., \text{ where } \lambda_{elem} = 1 - \mathrm{NUT} + \mathrm{NUT}^2 + ... \; .$$

The convexity coefficients λ_{elem} and λ drift *in a similar fashion* for large F towards the same value, *i.e.*, 1.

The difference between the two approaches is more evident in the transient state than in the steady state. In particular, the "finite system response" aspect is hidden in the single compartment model, which can be detrimental from a control perspective.

The outcome of this process is a dynamic model that is capable of controlling the output of the heat transfer fluid and its thermal profile under variable conditions, *e.g.*, the product temperature as it traverses the heat-treatment oven.

10.2.4 Counter-current Exchanger

The counter-current exchanger process may be represented in the form of elementary volumes (see Figure 10.8). An exchanger consists of two volumes V_p and V_f the volumes of the product and fluid, respectively. The corresponding volumetric flow rates are given by Q_p and Q_f. The heat-transfer surface and the coefficient are represented by A and U, respectively.

The steady state product output temperature is given by a convexity relation of the form:

$$T_{sp}(n) = \Gamma\left(Q_f, Q_p\right) T_{ef} + \left(1 - \Gamma\left(Q_f, Q_p\right)\right) T_{ep} \; .$$

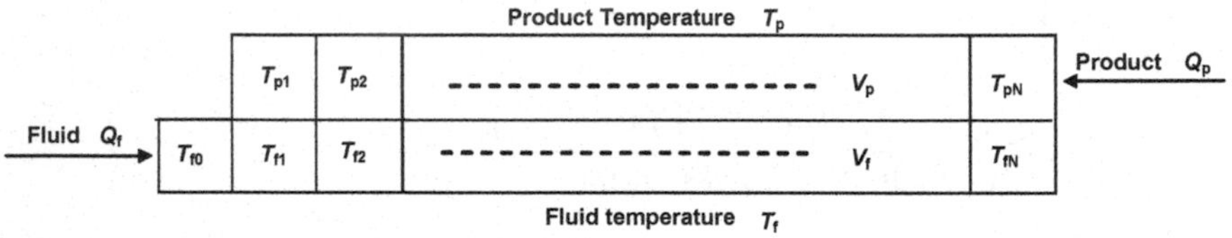

Figure 10.8. Counter-current exchanger

A summary of the calculation of $\Gamma\left(Q_f, Q_p\right)$ is presented. It is unusual in that the product compartmental analysis proceeds in the direction opposing the heat-transfer fluid outflow (where T_{p1} is assumed known and a value for T_{pN} is being sought). However, the direction of the heat-transfer fluid outflow is followed when analysing the heat-transfer fluid compartments, *i.e.*, from T_{f0} to T_{fN} .

Procedure:

- The exchanger is divided into N segments.
- The heat-balance equations between these segments are then stated, starting from the initial conditions of the known measured value of the heat-transfer fluid T_{ef}, referred to as T_{f0}, and ending at the assumed temperature value of the product output Q_1.
- The elementary heat-transfer function relationships between the fluid and product are expressed in the form of a matrix.
- This matrix is diagonalised and raised to the power of N to determine the final values of T_{pN} and T_{fN} using the identity $\underset{N\to\infty}{\text{Lim}}\left(1+\dfrac{Z}{N}\right)^N \cong e^Z$. Thus, T_{pN} and T_{fN} are connected to T_{ef} and T_{p1} by a static convex equation.
- Using the convex equation, T_{p1} is extracted as a function of T_{ef} and T_{pN}, where, in this case, $T_{p1} = T_{sp}$ and $T_{pN} = T_{ep}$.
- $T_{sp} = T_{p1}$ is expressed as a steady state, convex relationship where the convexity coefficient is a function of the Q_f and Q_p flow rates and the number of transfer units (NUT).

This procedure presents a clear view of the heat transfer within the exchanger. The relationship, obtained and verified by experience, is valid regardless of the relative heat flow rate values of the fluid and product (ρCpQ).

10.2.5 Implementation of a First-principles Model Controller

The classical implementation of a predictive controller consists of several stages, of which the first and last are the most complex.

10.2.5.1 Identification
Some problems may arise if the model used has been identified via the application of test-sequence protocols. Plant managers will insist that any disruption to production must be kept to a minimum. Consequently, managers will tolerate the application of small-amplitude test signals to their process for short periods of time only.

10.2.5.2 Implementation
Once determined, the final control solution must be implemented by computer. The ideal approach, from a software perspective, would be to program the solution using an internationally certified library of generic control modules that may be customised for the particular problem at hand. Such a procedure would minimise any potential reliability issues that might arise from custom-written software and would ultimately prove more economical.

10.2.5.3 How Are These Requirements Satisfied?
A knowledge-based or first-principles model, if one is available, may be used to reduce or eliminate the need for the application of test sequence protocols. A certain degree of effort is required to determine a knowledge-based model. However, once a model has been obtained and validated, it may be stored and re-used.

The proposed predictive control design may be expressed in the form of a library of graphical blocks that may also contain the internal knowledge-based model block. The final solution would be equivalent to graphically connecting the appropriate control blocks and inserting the specific numeric process and control values into the associated system diagram components.

In the case of the exchanger, inserting the particular device's physical parameters would be sufficient to allow the calculation of a dynamic, real time, internal model using a PLC. These parameters would constitute the physical characteristics of the fluids, *e.g.*, temperatures, flow rates and modes of operation.

As will be shown, this procedure will result in a regulator that will adapt to the various modes of operation that may result from the known variations of the physical parameters of the process.

10.3 Institut de Régulation d'Arles Exchanger

The Institut de Régulation d'Arles (IRA) is an organisation that has been operating and teaching industrial control systems since 1967. The facilities at the Institute are ideal for comparing the results of individual control techniques objectively.

10.3.1 Heat-exchanger Model

A physical representation of the exchanger is shown in Figure 10.9. In this case the product is cold water and the heat-transfer fluid is hot water. The flow rates of cold water Q_p and heat-transfer fluid Q_f are adjusted using PI regulators that compensate for the usual shortcomings of the valves. The fluids are at atmospheric pressure, and the temperature and flow rate magnitudes are as indicated in Figure 10.9.

A steady state model, determined previously (see Appendix B.1), is used to calculate the heat-transfer coefficient U, which for a given configuration, depends on the temperature and flow rate of the fluids.

The variation of U may be expressed as a function of the Reynolds, Prandtl and Nusselt numbers. These numbers depend on the conductivity, specific heat capacity, density and dynamic viscosity of the fluids, which are functions of temperature and flow rate. These classic functions are tabulated in the literature (see Appendix B.1). They may be represented by fouth order polynomials, which are also given in Appendix B.1. These simple algebraic relationships may be implemented in a straightforward manner in the control PLC.

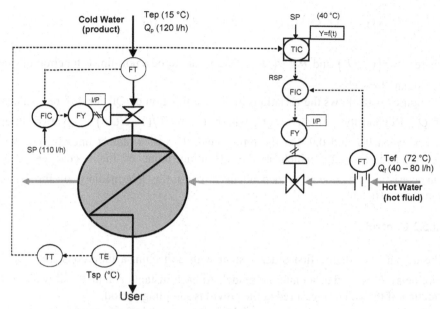

Figure 10.9. Physical diagram of the exchanger

For a particular flow rate Q and temperature T, which is variable for the given region, a combined "black-box model of a first-principles model" (referred to as the "BB/FP" procedure)   may be used to represent the coefficient U by a polynomial function of the form $U(Q, T) = (k0 + k1.T)Q^{0.8}$. The coefficients k_i may be estimated off-line using the calculated variable U. Experience has shown that the system dynamics may be approximated adequately by a first-order system with a pure time delay.

The pure time delay is represented as a linear function of the product flow rate (cold water), which is taken as the transportation time of the product through the exchanger.

It is necessary to perform a validation test on the knowledge-based model to recalibrate the heat-transfer coefficient that may otherwise drift. The Reynolds number calculation depends, in a complex fashion, on the physical configuration of the exchanger. The temperature of the fluid may be used to calculate the heat-transfer coefficient U that, for simplicity, is assumed equal to the average of the input and output temperatures. Such measurements are rare in industry and may be calculated using the model during the previous sample period. T_{sp}, the product output temperature, is the desired solution of the equation:

$$\tau \frac{dT_{sp}}{dt} + T_{sp} = T_{eq}(t - \theta),$$ where $T_{eq}(t)$ is referred to as the *equivalent temperature* and plays an important role.

$T_{eq}(t)$ is expressed, using the convex relationship between the input fluid temperature T_{ef} and the product temperature T_{ep}, as:

$$T_{eq}(t) = \Gamma\left(Q_f, Q_p\right) T_{ef}(t) + \left[1 - \Gamma\left(Q_f, Q_p\right)\right] T_{ep}(t)$$

where $\tau = \tau\left(Q_f, Q_p\right)$ and $\theta = \theta\left(Q_p\right)$. T_{eq} appears to be a non-linear function of the manipulated variable.

Figure 10.10 shows the variations of Γ as a function of Qf for different values of Q_p. Beyond the point $Q_f = Q_p$, where $\Gamma = NUT/(1 + NUT)$, Γ varies little, *i.e.*, between 0.7 and 0.9. On the other hand, Γ varies almost linearly between $Q_f = 0$ and $Q_f = Q_p$, which is the operating point of the exchanger. This characteristic is temperature dependent through the intermediary of the heat-transfer coefficient.

10.3.2 Control

The output T_{spm} of the first-order system with unit gain, a time constant τ and pure delay θ is used as an internal model. At each instant, τ and θ will vary as a function of the values measured at the previous sampling period.

In this case, the manipulated variable, *i.e.*, the equivalent temperature T_{eq}, is taken as the input to the model. This gives a classical control problem (see Chapter 4). The modelled output of T_{sp} and its predicted output after r samples are denoted by T_{spm} and T_{spred}, respectively. This gives:

$$T_{spm}(n) = a_m T_{spm}(n-1) + b_m T_{eq}(n-1), \text{ and } T_{spred} = T_{sp}(n) + T_{spm}(n) = T_{spm}(n-r),$$

where $r = \text{round}\left(\dfrac{\theta}{T_{ech}}\right)$. The equivalent manipulated variable is given by:

$$T_{eq} = \frac{\left(\text{Cons} - T_{sp}(n)\right) l_h + b_m T_{spm}(n)}{b_m},$$

where $a_m = e^{-T_{ech}/\tau}$, $b_m = 1 - a_m$, $l_h = 1 - e^{-3T_{ech}/TRBF}$ have their usual meanings.

The accuracy and robustness of the controller may be increased by using a second-order internal model that will take the dynamics of the regulated flow rate into account.

The next objective is to regulate Q_f using T_{eq} and to determine the set-point of the level 0, PI controller (flow regulator). T_{eq} is extracted from the equation:

$$T_{eq}(t) = \Gamma.T_{ef}(n) + (1 - \Gamma) T_{ep}(n)$$

from which the solution of the parameter Γ_s is given by: $\Gamma_s = \dfrac{T_{eq}(n) - T_{ep}(n)}{T_{ef}(n) - T_{ep}(n)}$,

where T_{ef} and T_{ep} are measured quantities that are assumed to be known.

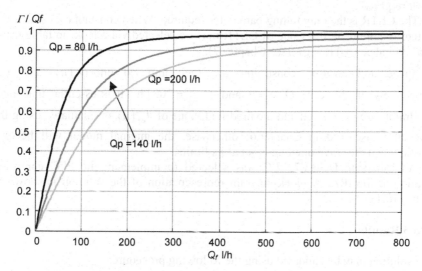

Figure 10.10. Variation of the convexity coefficient Γ as a function of the Q_f and Q_p flow rates

Otherwise, Γ_s and Q_f are connected by the non-linear, but one-to-one mapping function (see Figure 10.10), where Q_p and all temperatures are determined using the global heat-transfer coefficient U as an intermediary. This gives:

$$\Gamma(Q_f) = \frac{1 - e^{-UA\left(\frac{1}{F_p} - \frac{1}{F_f}\right)}}{1 - \dfrac{F_p}{F_f} e^{-UA\left(\frac{1}{F_p} - \frac{1}{F_f}\right)}}.$$

The heat flow rates are given by: $F_p = (\rho Cp)_p Q_p$ and $F_f = (\rho Cp)_f Q_f$. The global heat-transfer coefficient U is calculated at each sample point using:

$$\frac{1}{U} = \frac{1}{U_f} + \frac{1}{U_p}, \text{ where } U_i \text{ are the heat-transfer coefficients of each}$$

compartment. Thus,

$$U = U\big(Q_f(n-1),\ Q_p(n-1),\ T_{ep}(n-1),\ T_{sp}(n-1),\ T_{ef}(n-1),\ T_{sf}(n-1)\big).$$

The proposed solution involves determining Q_f from Γ_s using the classical iterative technique with linear interpolation as the last iteration stage. The maximum number of iterations possible between sample points is determined by the real time computational capacity of the PLC in question and the accuracy of the result required.

The CLTR is the only tuning parameter required. When constant robustness is a system requirement, the value of CLTR may be selected via a look-up table using τ as an index CLTR $= g\tau$ (see Chapter 7).

Speed and position constraints are then imposed on $Q_f(n)$ and Q_f, respectively. In this case $Q_{fmin} = 0$ and $Q_{fmax} = 200\,l/h$. It is essential that these constraints are transferred and the modified value of $T_{eq}(n)$ is calculated using the procedure discussed in Chapter 6; otherwise, the internal model will become invalid when the valve opening is set at its limits.

An Emerson, DeltaV7 PLC was selected to implement the PFC controller solution at the IRA. A block-diagram representation of the solution is shown in Figure 10.11.

10.3.3 Results

The solution may be validated using the following procedure:

- Verify the quality of the internal model prediction.
- Perform various control tests for different product flow rates and input heat-transfer fluid temperatures.
- Compare these results with those of a PID regulator, controlled by a third party, for different product flow rates using the following parameters, which result in tight regulation: $G_p = 2.8$, $T_{rep} = 17\,s$.

Also, the temperature T_{sp} is measured by a sensor situated downstream at the pipe outlet. This situation is common in practice and introduces a supplementary pure time delay. The results of the PFC regulator are then compared to those of the PID regulator.

10.3.3.1 Control Test
Figure 10.12 shows the effects of two product temperature set-point changes and a large change of product flow Q_p, *i.e.*, from 175 l/h to 75 l/h. These changes cause an immediate variation in heat-transfer fluid flow Q_f from 200 l/h to 80 l/h. This results in only a small variation, *i.e.*, 1 °C of the T_{sp} product temperature. Under the same conditions, a PID controller would only react *a posteriori* in response to the temperature variation.

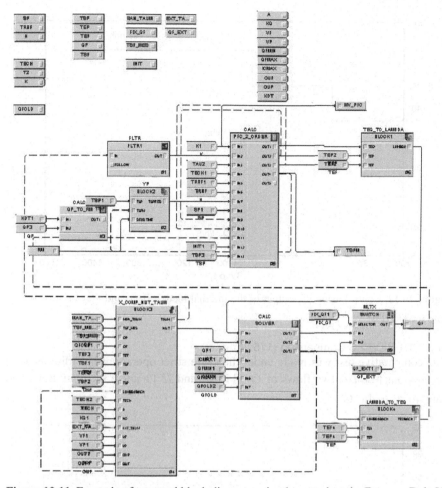

Figure 10.11. Example of a control block diagram as implemented on the Emerson DeltaV7

It is worth noting that we are not dealing with a classical feedforward *state variable* procedure such as a PID in a classical scheme, referred to as "PID FF" (or feedforward PID). In fact, it is a physical self-adjustment of the controller *structure* because Q_p does not act as an additive disturbance but as a modification of a structural parameter; it changes the effectiveness of the manipulated variable that has a very significant impact on the control.

The output of the T_{spmph} model follows the temperature T_{sp} adequately and only diverges by 3 °C during the transition resulting from the change of product flowrate Q_p.

Figure 10.13 shows the effect of a measured variation in the temperature T_{ep} from 56 °C to 45 °C that, despite precipitating an instantaneous change in Q_f, results in only a minimal error in T_{tsp} of 0.4 °C.

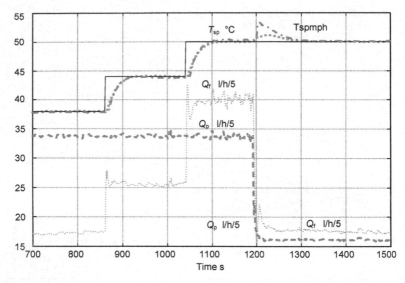

Figure 10.12. Control using the output of the internal model T_{spmph} and the variation of Q_p expressed as $(l/h)/5$

10.3.3.1.1 PID/PFC Comparison (175 l/h)

This comparative test was made under the same strict operational conditions as in the previous test, *i.e.*, two flows of Q_p, 175 l/h and 75 l/h and a set-point transition from 40 °C to 45 °C.

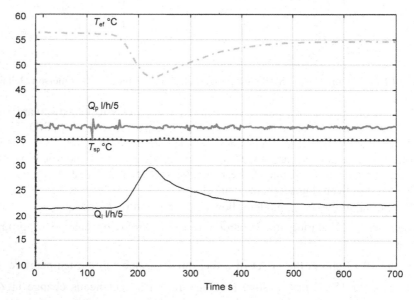

Figure 10.13. Variation of the input product temperature T_{ep}; Q_l fluid expressed in $l/h/5$

The response times of the two regulators were recorded as shown in Figure 10.14. It can be seen that the PID regulator, "tightly" controlled by an experienced independent human operator, is close to instability. Its 95% response time is approximately 82 s; whereas the predictive regulator produces a response time of 36 s. Note, the behaviour of the Q_f (MV) in each case is very different.

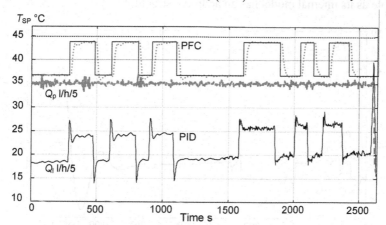

Figure 10.14. Comparison of PID and PFC (175 l/h) control

The initial high-frequency component of MV of the PID regulator has no effect on the output, while the equivalent PFC regulator MV is smoother. A sensor is placed at a distance from the process, resulting in the introduction of a time delay (or transportation lag). Figure 10.15 shows that the resulting behaviour of the PID regulator is marginally stable. On the other hand, the behaviour of the PFC does not vary since the time delay is accounted for in the internal model.

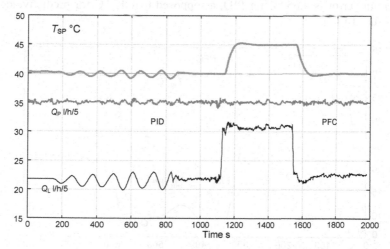

Figure 10.15. Comparison of the PID and PFC control with pure delay (175 l/h)

10.3.3.1.2 PID/PFC Comparison (80 l/h)

The operation at 80 l/h is more delicate because the local gain is larger and the process responds more slowly. It can be observed in Figure 10.16 that for the same tuning the PID regulator with a constant set-point is strongly unstable and divergent when $t > 1000$. However, the operation of the predictive regulator is acceptable as its internal model has an adaptive structure.

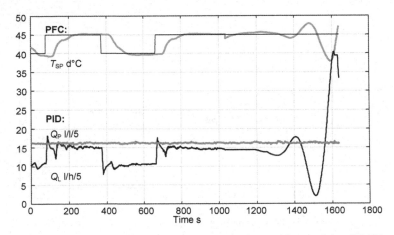

Figure 10.16. Comparison of a PID and PFC regulator with pure delay (80 l/h)

10.3.3.1.3 PID Control for a Flow rate Change Q_p

The same conditions as shown in Figure 10.16 are assumed for this test. A large change in the flow rate of Q_p occurs, *i.e.*, from 170 l/h to 80 l/h. The fluid flow rate reacts solely in response to changes in the temperature T_{sp}. The maximum temperature error is 4.65 °C for PID, as opposed to 1.15 °C for predictive control (see Figure 10.17).

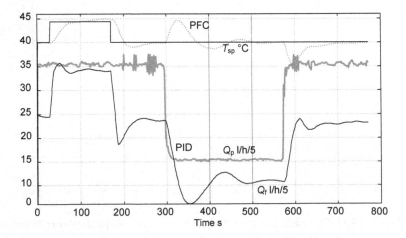

Figure 10.17. PID control in response to a flow rate change in Q_p

However, recall that the objective of this study is to show the benefits of the introduction of knowledge-based model control in the form of generic blocks. Its objective is not to compare the PFC and PID regulators, which have their own particular application domains.

10.3.3.2 Conclusions

The heat exchanger may be approximated by a first-order process with a pure time delay.

If the product flow rate, input product temperature and heat-transfer fluid temperature were constant, a PID regulator would produce satisfactory results because the need for dynamic regulation would be minimal.

The variations of product flow rate and fluid temperature behave as structural disturbances that vary the gain and dynamics of the transfer function, *i.e.*, the manipulated variable is the fluid flow rate $MV=Q_{\text{fluid}}$, the regulated variable is the CV = exchanger output temperature and the product temperature is an additive state disturbance. If changes occur in Q_p, T_{ep} and T_{ef} it is advisable to vary the parameters of the model as a function of these three variables. Some work would be involved if a representational model were used; but no such modifications would be required in the case of a first-principles model.

It may be argued that it is also theoretically possible to change each tuning parameter of a PID regulator in accordance with these three variables. However, this would still require considerable effort.

In addition to its ease of tuning, the predictive regulator is more effective because it is adaptive with an internal, first-principles model implemented in a generic control library.

The above conclusions would seem to imply that PID regulators should only be used when environmental conditions remain constant, and that model-based controllers should be reserved solely for variable environmental conditions. However, if the system model has been determined and the PLC control library contains a first-principles model block, the overhead involved on the part of the installer in using such a block is negligible.

There are a number of benefits associated with having to choose only a single parameter with a clear, physical interpretation, *i.e.*, the closed-loop time response. The simplifications in tuning would drastically reduce tuning times, result in better performance and permit a more flexible maintenance scheme to be employed.

10.4 ARCELOR

The following applications have been implemented at the SOLLAC steel mill (ARCELORMITTAL group) in Fos-sur Mer (France).

10.4.1 Continuous Casting

10.4.1.1 Process Description

The liquid steel arrives at the plant in batches of 335 tonnes at an approximate temperature of 1550 °C (see Figure 10.18). Each batch is delivered at the top of the machine via a carousel that is equipped with a drainage valve referred to as the rotary nozzle. The steel flows continuously into a 60-tonne reservoir called the tundish. The tundish possesses a nozzle that supplies liquid steel to the mould. The nozzle opening consists of 2 or 4 slits. The liquid steel flow rate through the nozzle is controlled at its input by adjusting the position of the tip of the stopper.

The ingot mould is rectangular in shape and one meter in length. The width of the mould is variable and its copper walls are actively cooled using water to promote surface hardening. While supporting the hardened skin, the draw rollers permit the product to be pulled out at a rate of 1 to 1.8 m/min. Each section is cut to length using blowtorches. The finished product is referred to as an ingot (of thickness 220 mm, width 850 to 2200 mm and length 5.8 to 14.6 m). The ingot is then transferred to an annealing furnace and ultimately to a hot-rolling mill.

10.4.1.2 Control Loops

The weight of the steel in the tundish is regulated by a controller whose function is to ensure a relatively constant level while taking any MV constraints into account, *i.e.*, the rotary nozzle. This actuator is a fragile and expensive mechanism consisting of two superimposed, heat-resistant plates; one stationary and the other mobile. Each plate possesses a circular hole; the flow control is achieved by altering the relative positions of the mobile opening. It is important to minimise any rotational movements of the mobile plate in order to limit wear and increase the actuator life span.

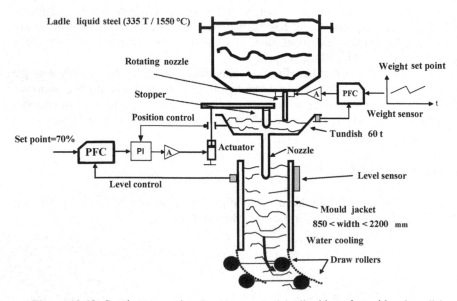

Figure 10.18. Continuous casting. Level control of the liquid-steel mould and tundish

The mould oscillates with an amplitude of 8 mm peak to peak and an approximate frequency of 2 Hz. The level of the mould is regulated by a second PFC controller that is used to adjust the position of the stopper. This configuration results in a process that is difficult to control. The supply nozzle is immersed in the liquid steel and the angular orientation of the jets at the slit outputs is such as to generate a mixing motion that improves uniformity.

The steel surface is permanently blanketed with a fissile powder to protect the steel against oxidisation and ensures that sufficient lubrication exists between the liquid steel and copper face of the mould. The internal mixing of steel induces surface waves.

Aluminium aggregate disrupts the liquid steel flow; this may be eliminated by injecting argon at the stopper tip. The degree of stability dictates the quality of the finished product: level variations disturb the blanket of powder particles on the surface, giving rise to inclusions at the centre of the ingot, thus making it unfit for lamination. As a first approximation, a specification of less than ±5 mm peak to peak is selected.

10.4.1.3 Parameters

- CV: Sensor measuring the degree of ingot hardness, based on the absorption of γ – radiation by the iron: $[0-100]$ mm.
- MV: PI control of the stopper position $[0-160]$ mm.
- DV: Extraction speed of 1 to 2 m/min.

The model – The general behaviour of the mould for a typical ingot may be represented by a second-order system with a short time delay in the region of 0.1 s, which can be disregarded.

$$\frac{\Delta H\left(\%\right)}{\Delta X\left(\text{mm}\right)} = \frac{b_0 e^{-\theta s}}{s\left(1 + a_1 s + a_2 s^2\right)} .$$

10.4.1.4 Identification of the Transfer-function Parameters

Some results of the procedure applied to the mould are now presented to illustrate its operation. The procedure may be divided into three steps:

- test sequence protocol application;
- local identification;
- global identification.

10.4.1.4.1 Mould Test Sequence Protocol

The mould test-sequence protocol was designed to minimise the uncertainty of the process parameters estimation. A deterministic protocol (as opposed to a stochastic approach, e.g., pseudo-random binary sequence) was used. The supervisor of the mould process imposed stringent restrictions on the tests that may be carried out on the plant. The two parameters affected were:

- the time available for the application of the test sequence protocol was limited;
- the test signal itself had a limited amplitude $\pm A$ to minimise the impact on normal production.

Thus, the only degree of freedom available for optimisation was the sequence of step durations. The applied sequence contained a step long enough to sensitise the steady-state gain. It also contained a period sufficiently fast to sensitise the dynamics in order to minimise the variance of the estimated process parameters.

In general, any of the standard protocol optimisation procedures may be used, provided a tentative model of the process is available! But, regardless of the procedure adopted, it is important to avoid becoming trapped in a vicious cycle of "I can optimise the protocol if I know the process". The test sequence applied, in the case of the moulding process, is shown in Figure 10.19.

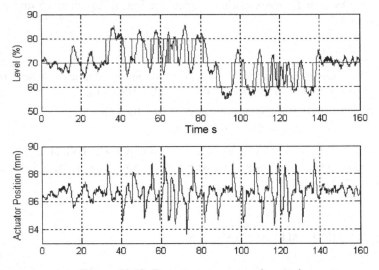

Figure 10.19. Test-sequence protocol example

10.4.1.4.2 Local Identification

Figures 10.20 and 10.21 demonstrate a sample of the results obtained from the application of the test-sequence protocol to the moulding process. Identification was achieved using the results from several tests in closed loop.

A parallel filtering technique was employed to pre-process the input and output signals. A low-pass filter with $T_{lp} = 1$ s was used to eliminate measurement noise and a high-pass filter $T_{hp} = 10$ s , respectively was applied to eliminate the effects of drift. The transfer function was then estimated using the Gauss – Newton algorithm for parameter estimation, resulting in the expression:

$$\frac{\Delta H}{\Delta X} = \frac{3.3}{s\left(1 + 0.497s + 0.404s^2\right)} .$$

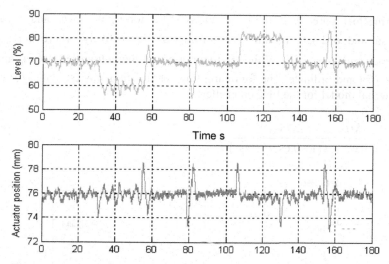

Figure 10.20. Process response to first application of test-sequence protocol

The criteria for the model absolute error value is: C% = 16.36. For the second test, a transfer function of the following form was determined:

$$\frac{\Delta H}{\Delta X} = \frac{3.77}{s\left(1+0.579s+0.390s^2\right)} \ .$$

The criteria for the model absolute error value is: C% = 17.62 .

10.4.1.4.3 Global Identification

The intervals of uncertainty on each of the parameters b_0, a_1 and a_2 were calculated for a level of iso-criteria of 10%. Figure 10.22 illustrates the resulting iso-distances.

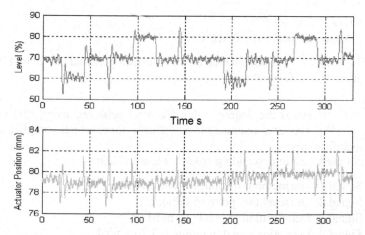

Figure 10.21. Process response to the second application of the test-sequence protocol

Referring to Figure 10.22, the two regions at the top-right use b_0 and a_1 as the abscissa and ordinate (x and y co-ordinates), respectively where $num_0 = b_0$, $den_1 = a_1$ and $den_2 = a_2$. While the two regions at the bottom-left use a_1 and a_2 as abscissa and ordinate. The following transfer function was estimated for use as the internal model for the PFC controller:

$$\frac{\Delta H}{\Delta X} = \frac{3.75e^{-0.1s}}{s(1+0.6s+0.34s^2)} \; .$$

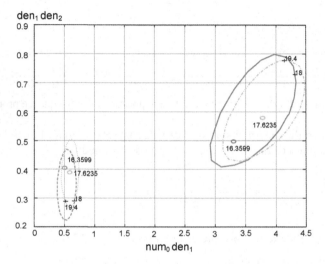

Figure 10.22. Iso-distance curves

PFC Parameters:

- T_s = 46.88 ms.
- CLTR = 2 s.
- Coincidence horizon = 0.516 s.

The PFC controller was implemented on an industrial personal computer (IPC) and was programmed in Turbo Pascal.

10.4.1.4.4 Results
Figure 10.23 compares the degree of level control achieved using PID and PFC control. For PFC the level remained inside a ±5% zone, which complies with the system specifications.

Signals presented in descending order in Figure 10.23:

- Signal 1: The method of control chosen (PID or PFC).
- Signal 2: The level (set-point of 70%).
- Signal 3: Jack position in mm (~88 mm).
- Signal 4: Extraction speed in m/min (~1.3 m/min) .

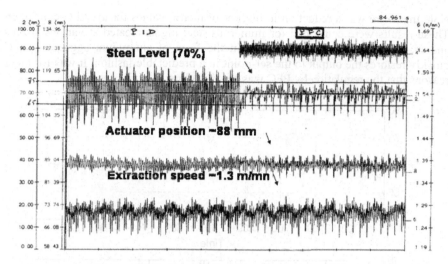

Figure 10.23. Comparison of PFC and PID regulation

10.4.1.4.5 Histograms

In order to compare the quality of level control achieved by the PFC and PID control, the level error (Setpoint – measured output) has been determined in each of the two cases. Figures 10.24 and 10.25 demonstrate the regulation errors obtained in the PFC and PID cases, respectively. The errors are presented in histogram form as the percentage occurrences of the resulting average and standard deviation errors recorded during the tests.

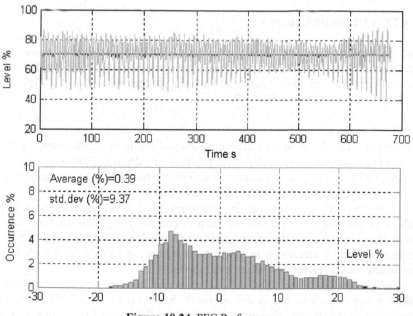

Figure 10.24. PFC Performance

Note that there was a reduction in both error metric values for the PFC controller. These results were confirmed for numerous steel ingots created at various speeds and mould widths. For disturbances, in the form of aluminium aggregate, PFC quickly restores the output to the set-point in a predictable fashion. It was for these reasons that the regulation by PFC was readily adopted by the operators.

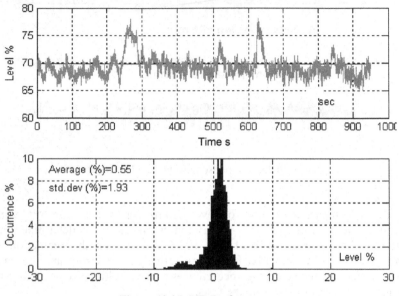

Figure 10.25. PID Performance

10.4.2 GV1/2 Steam Generators

The Fos factory has four steam generators (GV) that are used to power turbo-compressors and alternators. The furnaces are capable of burning multiple fuels and can produce 215 t/h of steam at 80 bar superheated to 500 °C. The output power of each generator is 192.2 MW.

10.4.2.1 Introduction

The furnaces can function with different combustible fuels: blast-furnace-enriched gas (Ghfe), coke-furnace gas (Gfc), heavy oil (Fo2), domestic oil, oily waters, coal tar (Goc) and natural gas. Generally, two fuels are burned simultaneously that are, for the most part, gaseous (Ghfe/Gfc) (Figure 10.26).

During normal operation the furnaces are subject to frequent changes in fuel, each of which possess different combustion characteristics (steam flow rate and temperature) imposed by the plant operation. Major load variations (*e.g.*, ±50 t/h in 3 min.) produce fluctuations in steam temperature and pressure, which in turn, have a detrimental effect on the output of the turbines.

It is the change of fuels and the uncertain and frequent variations in load that make the use of a classical control approach, based on optimised PID controllers, unsatisfactory. Consequently, several internal model-based PFC controllers were substituted for the original PID controller in order to increase the static and dynamic performance of the principal control loops (pressure and temperature); while reducing the operators' workload.

The overall improvement in the control of the plant permitted further set-point refinements to be introduced resulting in a subsequent increase in energy output.

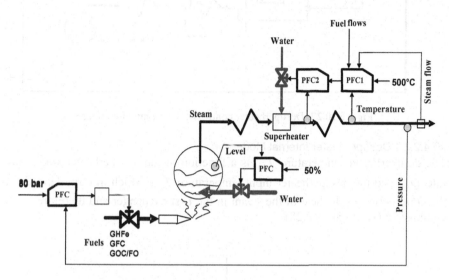

Figure 10.26. Steam generator

10.4.2.1.1 Control Implementation
The level 0 control was implemented using a Schneider Electric PMX7 PLC, *e.g.*, control of the desuperheater water flow rate. The higher-level control was also implemented in a Schneider Electric PLC that included implementations of various PFC predictive controllers in its library.

10.4.2.2 Superheater
The output of the steam ball is superheated by 2 exchangers driven by the exhaust fumes of the combustion. A desuperheater inserted between the exchangers, denoted E3 in the figure, ensures the control of the steam temperature by injection of water (see Figure 10.27).

The control of the superheater-desuperheater is achieved by 2 PFC in cascade (see Section 7.3), the first controller regulates the superheater output temperature T_{des} and the second controls the output temperature of the second exchanger T_{vap}, this implies the temperature of the used steam supplies the turbines.

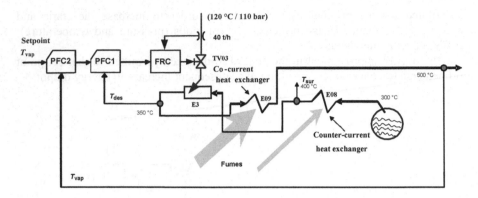

Figure 10.27. Desuperheater control of the steam temperature

10.4.2.2.1 Desuperheater Internal Model

The desuperheater temperature T_{des} is a function of the desuperheater water flow rate Q_{des} and the desuperheater input temperature T_{sur}, which in turn, depends on the contribution of the heat of the steam in the first exchanger and therefore to the combustion (see Figure 10.28).

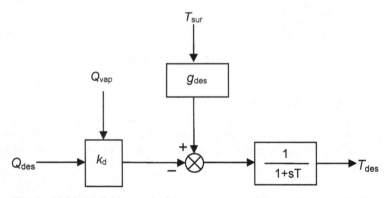

Figure 10.28. Model internal of the temperature of desuperheater controller

10.4.2.2.2 Desuperheater Identification

The test-sequence protocol was superimposed on the desuperheater valve control signal. The desuperheater output temperature is influenced by the desuperheater flow while taking the steam flow rate and superheated steam temperature into account. Figure 10.29 shows both physical and simulation results obtained for the applied test-sequence protocol.

The identified parameters were g_{des} (a value near d_{e1}), a time constant T (approximately 25 s) and the gain k_d. The value of k_d decreased as the Q_{vap} load increased. The variation in kd took the form of a quadratic, i.e.,
$$k_d = a + bQ_{vap} + cQ_{vap}^2.$$

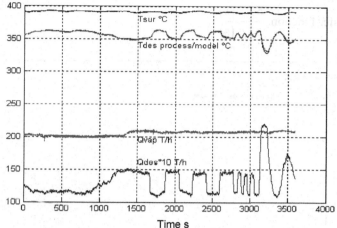

Time s

Figure 10.29. Test-sequence protocol $\left(Q_{\text{des}}, Q_{\text{vap}}, T_{\text{sur}}/T_{\text{des}}\right)$

10.4.2.2.3 Superheater Internal Model

Figure 10.30 shows how the dynamics of the output temperature of the second exchanger T_{vap} may be explained when the collective influence of the temperature T_{des}, the load Q_{vap} and the heat contribution of the steam Q_{thfum} are taken into account.

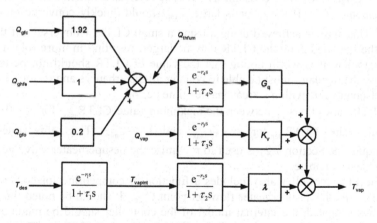

Figure 10.30. Internal model of the steam temperature controller

10.4.2.2.4 Steam Temperature Identification

Figure 10.31 shows the data used in the identification of the input temperature of the desuperheater and output temperature of the second exchanger, T_{des} and T_{vap}, respectively. In Figure 10.32 a cascade control configuration was adopted (see Section 7.3).

PFC2 uses a third-order system process as its internal model. The model consists of a first-order system, representing the transfer function of the desuperheater temperature regulation T_{des} in closed loop, followed by the second-order transfer function $\dfrac{\Delta T_{\text{vap}}\left(s\right)}{\Delta T_{\text{des}}\left(s\right)}$.

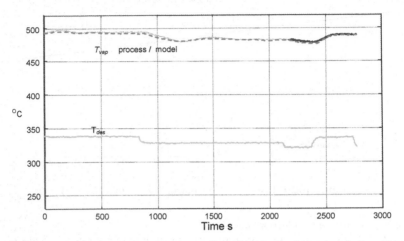

Figure 10.31. Test sequence protocol used to identify the steam temperature T_{vap}

In this case, the PFC controller utilises zone control (see Section 7.2). This implies that the closed-loop time constant CLTR is selected as a function of the error, *i.e.*, $\varepsilon = \text{setpoint} - T_{\text{vap}}$. If the error is large, T_{vap} should quickly converge on the set-point. This may be achieved using a using a small CLTR. Conversely, if the error is in the specified zone the CLTR may be larger, resulting in more robust control configuration. It is worth noting that the value of CLTR should not be too large because the regulator must be able to counter all disturbances and restore the CV to the set-point value within a "reasonable" time period.

CLTR varies linearly between the maximum value CLTR_{max} for $\varepsilon = 0$ and the extreme value CLTR_{min} for $\varepsilon > +\varepsilon_{\text{max}}$ and $\varepsilon < -\varepsilon_{\text{min}}$. Thus, this zone-control technique (see Section 7.2) is used to minimise the desuperheater valve activity in the presence of noise.

The PFC controller also includes a function to compensate for constraints on the MV, *i.e.*, the desuperheater flow set-point C_{qdes}. It should be noted that if C_{qdes} becomes saturated, the internal model of the controller should be made aware of this condition when it arises. Otherwise, the internal model risks becoming inconsistent and thus resulting in erroneous output predictions. The principle employed consists of supplying the internal model of the regulator with $C_{\text{qdes\_lim}}$, and not by the calculated MV, when $C_{\text{qdes}} = C_{\text{qdes\_lim}}$ (see Section 7.3.1).

Finally, this PFC controller compensates for the furnace and heat loads, Q_{vap} and Q_{thfum}, respectively (see Section 4.7). In addition, the internal model also includes two first-order models representing the disturbances present.

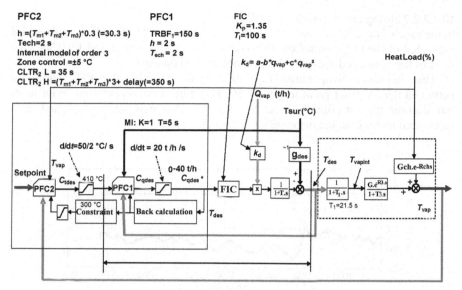

Figure 10.32. Controller overview for second exchanger output temperature T_{vap}

10.4.2.2.5 Performance of the PFC Control of T_{vap}

Figure 10.33 presents a histogram of the error (Setpoint − output error) of T_{vap} using PFC control. This may be contrasted with the histogram of a classical PID controller shown in Figure 10.34.

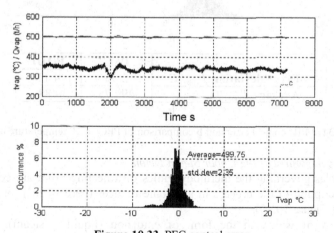

Figure 10.33. PFC control error

10.4.2.2.6 Histogram of the PFC Controller Error

This histogram illustrates errors in the range from − 10 to +10 °C only. The standard deviation of the error is 2.35 °C with an average output of 499.5 °C for a set-point of 500 °C. Note that the measured Tvap is centred on the 500 °C set-point. The steam load is 350/2 = 175 t/h. (Note the scale of the figure is divided by 2.)

10.4.2.2.7 Histogram of the PID Control Error

In the case of the PID controller, the error deviation is 4.11 °C, which is about 1.8 times that of the PFC controller. The average steam flow rate is 184 t/h. (Note the scale of the figure is divided by 2 and is not centred on zero.)

This improved temperature regulation achieved with the PFC controller permitted the Tvap set-point to set at 500 °C during normal operation. Prior to its introduction the set-point was set at 495 °C. This new setting resulted in a substantial increase in the turbo-generator outputs.

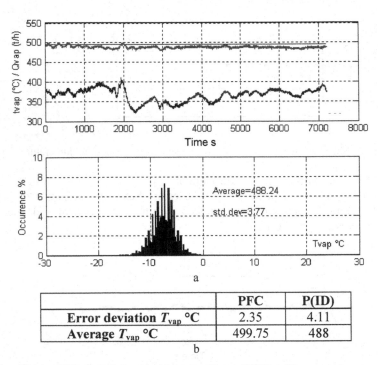

	PFC	P(ID)
Error deviation T_{vap} °C	2.35	4.11
Average T_{vap} °C	499.75	488

b

Figure 10.34. a PID control error and **b** comparison of PFC and P temperature controllers

10.4.2.3 Level Control of the Upper Steam Drum

The principles governing the thermodynamics of a heating drum are complex. A simple description of the process in a volume reduces (here 14 m³) for a flow of 230 t/h (see Figure 10.35).

The flow of water, in the form of emulsion (liquid + steam), from the evaporation tubes is equal to the water supply flow coming up from the lower steam drum.

A water flow rate is supplied under pressure at a temperature of 200 °C approximately, which avoids the quenching phenomenon.

The steam extraction flow rate is 84 bar at approximately 300 °C. The complex phenomenon of expansion/contraction that takes place in the evaporation tubes is also present in the ball and manifests itself as a rise in the level when an increase of the steam flow rate occurs.

The regulation of the fluid level in the heating drum is critical. If the ball were to run dry, this would quickly result in the destruction of the evaporation tubes. Conversely, an overflow in the heating drum would result in harmful liquid water entering the turbo-alternators. The flow required by the various steam consumers variables considerably and thus affects the level of the lower steam drum.

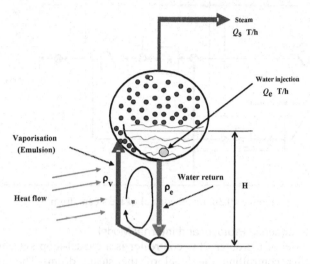

Figure 10.35. Steam drum principle of operation

10.4.2.3.1 Control Strategy

The control strategy includes a PID controller, in the PMX PLC, that receives the level set-point whose manipulated variable is given by:

$$MV = K\left(C_{\text{level}} - \text{Level}\right) + Q_{\text{vap}} \ .$$

The set-point of this regulator is the manipulated variable of a transparent regulator (see Section 7.4). The PID controller is a simple proportional P regulator that facilitates the easy transfer of constraints to the outer controller. It is worth noting that "back calculation" is not possible with a PI controller.

10.4.2.3.2 Feedforward Compensation

The flow of steam gives rise to the phenomenon of expansion/contraction that is taken into account by the transparent PFC controller (see Figure 10.36).

10.4.2.3.3 Zone Control

This procedure (see Section 7.2) is especially useful in this case because it produces "tight" control when the level approaches the high and low limits; whereas it can be smooth when the level is in the central zone of normal operation.

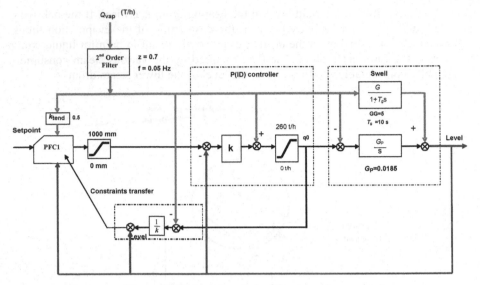

Figure 10.36. Level control of the steam drum

10.4.2.3.4 Test Sequence Protocol and Internal Model

The internal model of the outer PFC controller is a closed-loop system consisting of a P controller controlling the level of the steam drum. The input to the feedforward model (gain GG and time constant T_g) of the PFC controller is Q_{vap} (see Figure 10.37). The identification was made using the test sequence protocol shown in Figure 10.38.

- Maximum CLTR in the centre of the zone: CLTR = 350 s.
- Minimum CLTR outside of the zone: CLTR = 200 s.

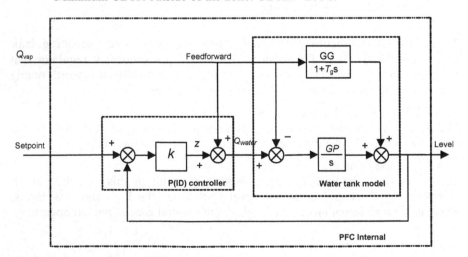

Figure 10.37. Overview of heating drum level control

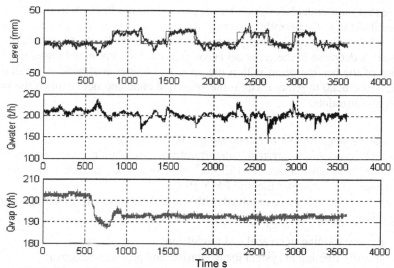

Figure 10.38. Parameter identification test sequence protocol PFC1 parameters for level control $T_s = 1$ s , $h_s = 1$ s

10.4.2.3.5 Measurement of the Level PFC Controller Performance

The difficulty in stabilising a loop with two integrators is well known. The difficulty associated with bringing a second integrator into a loop that already includes one integrator highlights a major weak point in the PI controller (level is the integral of the input and output flow rates difference).

The observed steady state error is due to the absence of the integral term, which has been removed due to the instability that would be introduced by its presence.

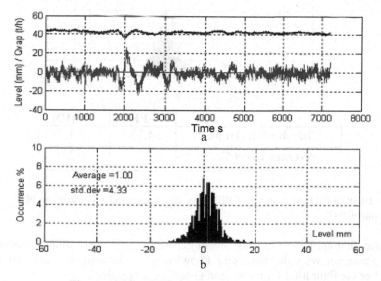

Figure 10.39. a PFC signal and **b** error histogram

Note that the standard deviation has decreased for the PFC controller. It should be noted that the PFC controller introduced a marked improvement in the system's ability to track the set-point during variations in the steam flow rate and changes in fuel used. A decrease in the error deviation is important, but maintaining the level set-point of the heating drum within the specified operational zone is paramount. Indeed, it is crucial to remain within the specified zone and not to touch the upper and lower constraints.

10.4.2.3.6 Training,Transfer of Expertise and Maintenance

All signal processing, system identification, simulation and control law development was carried out in MATAB® and Simulink® and the controller was implemented in a Schneider Electric PMX PLC.

The relevant employees, involved with control issues, were given appropriate training. This included:

- general training in control and modelling;
- specific training in system identification and predictive control;
- training in controller implementation on Momentum PLCs.

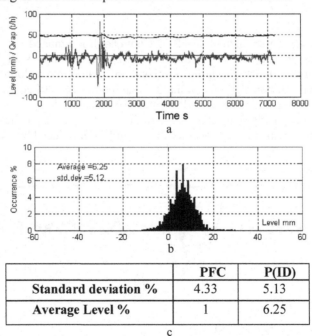

	PFC	P(ID)
Standard deviation %	4.33	5.13
Average Level %	1	6.25

c

Figure 10.40. a PID error signal, **b** PID error histogram and **c** comparison of PFC and P level controllers

PFC was first applied to the first steam generator. The staff training provided the plant personnel with the necessary knowledge to subsequently develop a PFC controller configuration for the second generator themselves.

This comprehensive approach to training will allow plant personnel to adapt the models and control configurations to deal with any variations in materials and specifications that may arise in the future (see Chapter 11).

10.4.2.4 Assessment

10.4.2.4.1 Economic Assessment: the "Squeeze and Shift" Procedure

Why is it of interest to reduce the variance of the steam temperature? The higher the steam temperature the higher the turbine efficiency, but if the temperature is too high due to the rotational speed and thermal expansion the blades of the input compressor start to "creep" and will ultimately hit the casing (see Figure 10.41). The procedure consists of:

- reducing the variance – "squeeze" (see Figure 10.42);
- moving the target to the upper limit – "shift" (see Figure 10.43).

In our case if you increase the steam temperature by 4 °C this results in a financial return of 1 million euro per year: in this case the pay out time (POT) is negligible.

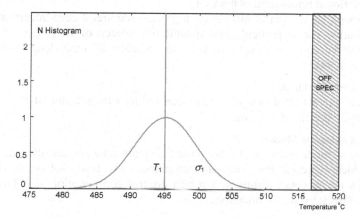

Figure 10.41. The initial distribution of steam temperature

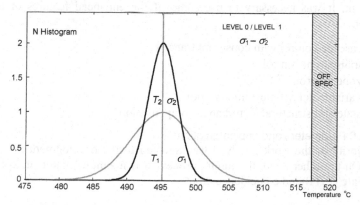

Figure 10.42. Improved dynamic control $\sigma_1 \rightarrow \sigma_2$ – "Squeeze"

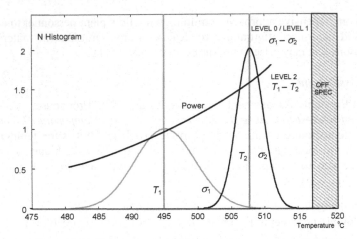

Figure 10.43. Constrained optimisation of the set-point – "Shift"

10.4.2.4.2 Better Knowledge of the Unit

The development of control models for a process requires a deep understanding of the qualitative and, in particular, the quantitative aspects of the process. All loops are interactive and if one were to become unstable, all other loops would also become unstable.

10.4.2.4.3 Performance

In general, the standard deviation has been reduced by a factor of two to three, depending in the circumstances.

10.4.2.4.4 Adaptive Model

The physical model varies as a function of the steam flow rate and the nature of the fuels, which represent the principal disturbances involved. Taking these model variations into account results in a robust control configuration – a fact that is greatly appreciated by the operators.

10.4.2.4.5 Controller Configuration

The results obtained could not be achieved using the predictive controller in its basic form. It was necessary to use some of the enhanced features of the PFC algorithm:

- cascade control with constraint transfer;
- transparent control;
- zone control;
- variable delay time compensation;
- state and structural disturbance compensation.

10.4.2.4.6 Parameters and Implementation

The majority of the work involved was devoted to the development of a process model. On the other hand, the tuning process was a simple matter of selecting an appropriate CLTR and coincidence point.

The software development was greatly facilitated by the fact that the predictive controllers are represented as standard graphical programming blocks in the Schneider Electric PLC software library.

10.4.2.4.7 Personal and Collective Effort

All the work required a collective desire to succeed. A large intellectual effort on behalf of those involved was required to analyse the different system subcomponents in order to generate a global simulator. Numerous iterations of the development process were required, *i.e.*, controller synthesis, test-sequence protocol generation, identification and simulation. The problems encountered during the development cycle were compounded by the constraints imposed by the production commitments of the plant.

10.5 EVONIK.DEGUSSA

EVONIK.DEGUSSA is the world number one in the area of specialty chemistry. Degussa was trained in PFC/PPC and adopted the technique in 2003. Since then they have implemented this technique in most of their domestic and international production sites. Many units, such as heat exchangers, dryers, generators, continuous and batch reactors, binary distillation columns, *etc.* are now controlled by PFC/PPC implemented in all kinds of PLCs and DCSs. First-principles/black-box modelling of the units were made in MATLAB® and stored in a library for general use. The implementation procedure, as well as the associated software, were standardised by the in-house control group in order to minimise the time and cost of implementation. The approach involved selecting one unit in the plant on a pilot basis, implementing the PFC/PPC control strategy and, when operating satisfactorily, deploying the final scheme to all the other units in the workshop.

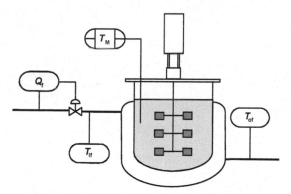

Figure 10.44. Batch reactor

This application involves a batch reactor (14 m³) operating in "lost-fluid mode" with no back circulation around the jacket. The reactant operational temperature plateau is reached by heating the jacket with a flow of steam. Reactants are introduced with the solvent. At a specific temperature an exothermic reaction appears and is regulated by a PI-controlled flow $(Q_f - \text{F.I.C.})$ of water at a variable, but measured, temperature. The PPC controller (see Chapter 8) insures that the temperature of the mass T_M remains constant (see Figure 10.44).

The flow of cooling fluid modifies the time constant and, since the exothermicity can vary to a large extent, the equivalent time constant varies accordingly. For a classical PI controller the process would appear quite non-linear and difficult to control. Since the accuracy required is very "tight" the control specifications would not be fully satisfied by conventional control.

A simple, first-principles model has been developed and validated by some experimental tests (see Figure 10.45).

A cascade of controllers (see Section 7.3) was implemented to control the batch-reactor temperature as shown in Figure 10.46. The purpose of the three controllers required in the process configuration is as follows:

- PI to control the flow of water (F.I.C.);
- PPC to control the mass temperature;
- PFC to remove the eventual bias coming from the exothermicity.

(See Section 6.4.)

The three controllers were implemented in the DCS Foxboro IAS (HLBL). A comparison was made between the former PI(D) control and the PPC controller (see Figures 10.47 and 10.48).

The achieved repeatability made it possible to optimise the operating set-point and thus increase the quality and quantity of the product. It is also important to mention that the operators appreciated the decrease in their workload brought about by the new controller. This procedure was deployed to all reactors as a result.

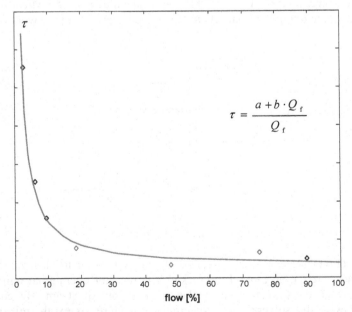

$$\tau = \frac{a + b \cdot Q_f}{Q_f}$$

Figure 10.45. Time constant as a function of the flow (normalised units)

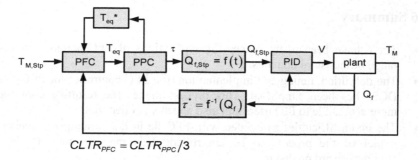

Figure 10.46. Cascade control of reactor

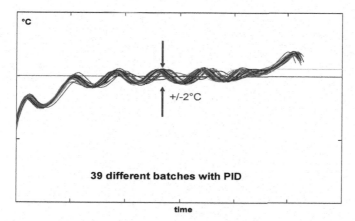

Figure 10.47. PI control

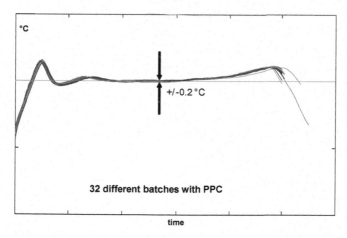

Figure 10.48. PPC control

10.6 Summary

- There have been many successful applications of PFC across a diverse range of industrial domains.
- The definition, tuning and implemention of a PFC controller for PLCs and DCSs has been simplified. This process makes the resulting controller more accessible to PID users and also appeals to operators.
- The main difficulties associated with PFC lie in determining the dynamic model of the process to be controlled. This may or may not be a straightforward procedure.
- The POT for a PFC installation is usually quite short.

11

Conclusions

Up to this point, only elementary problems have been considered. When the mathematical model of a process is available, it is possible to derive an equation for the controller. This approach may be described as homogeneous as the data and results are of a similar nature. Any rational solutions, which are mathematically feasible, may be solved with computer assistance.

However, obtaining a process model, which is the main tuning parameter of the controller, is another matter entirely. This latter approach is not considered homogeneous as the noisy and uncertain sensor data extracted from a complex real-world system is of a different nature from the precisely defined, mathematical formula of the controller.

Even with a good understanding of control theory, system identification and modelling should be approached with caution. What are the available choices? In any occupation, an expert possesses a comprehensive set of tools capable of solving a diverse range of problems. The tool that best fits the problem should be selected without prejudice.

Modelling requires the designer to delve deeply into the understanding of the process, whereas, in the case of PID, the selection of two coefficients suffices. Thus, PID represents a far simpler approach from a tuning perspective. This "simplicity" is gained at the cost of a reduction in control performance. Two approaches should not be used:

- Applying advanced control in all situations. The term "advanced" refers to any controller other than PID. This approach wrongly assumes that advanced control contains all the possible advantages of control techniques.
- PID "perseverance". Implementing a PID controller with the help of numerous ingenious tricks is quite satisfying for the designer. However, such an approach is hazardous and cannot be justified, as customised solutions are difficult to update and have poor re-usability potential.

PFC and PPC are open methods; a fact that has been clearly demonstrated in Chapter 7. Numerous academic works have contributed to the theoretical development of PFC and its associated techniques. This book has been motivated by those in industry who are interested in exploiting the potential benefits of this method of control.

11.1 Characteristics of PFC Control

Currently, a large number of PFC applications exists. What are the specific characteristics of PFC?

11.1.1 Model

The PFC controller uses an independent model. Any type of model may be used as long as it is capable of representing the process, *i.e.*, What will the response of the process be when subjected to a given input? It is not limited to the step response of a process, state-space representation, *etc*.

The *decomposition principle* opens up a large number of possibilities for transforming the model and ultimately combines the measured operation of the process with that of the prediction. Chapter 7 offers examples of this procedure.

The explicit distinction made between the free and forced outputs facilitates mixing the actions of various controller configurations, as shown in Chapter 7. This intuitive approach assists floor instrumentation personnel, accustomed to PID control, in understanding the concepts of predictive control.

11.1.2 Reference Trajectory

Considering the set-point as the only possible future reference of the process is to miss an opportunity to select the manner by which this set-point should be reached. The reference trajectory is as important as the set-point itself. Could there be anything more simple, or more intuitive, than selecting an exponential trajectory – or selecting a CLTR?

11.1.3 Regulator

The focus here is intentionally restricted to SISO processes (and also 2MV/2CV). This is done in order to maintain a clear objective, *i.e.*, to replace PID in an area where its efficiency is unsatisfactory. Predictive control offers, as an alternative to PID, a simple algebraic solver without iterative quadratic minimisation that functions efficiently and is easy to implement.

11.1.4 Basis Functions

The objective of the PFC controller is to produce an MV that minimises the energy of the deviations between the potential projections of the CV and the desired reference trajectory. Various projections of the future MV may result in similar behaviour of the CV. The design of the PFC controller can be simplified by restricting the selection of the potential future MV. In practice, a limited subset of the MV – CV basis functions is used to determine the most appropriate MV, *a priori*. An extra tuning parameter is included in the algorithm to control the degree of energy minimisation required.

As well as ease of tuning, the above strategy also has the benefit of following, for example, a polynomial set-point with zero tracking error when basis functions are chosen from the eigenfunctions of the process (*e.g.*, Taylor-series polynomial).

11.1.5 Time and Frequency Domains

Tuning may be carried out by taking the time specification (CLTR) and frequency specification (*i.e.*, sensitivity functions, dual control) into account. The strategy for using PFC is to maximize performance while maintaining ease of tuning and implementation. While maximising the theoretical performance of PFC is a challenging and interesting academic exercise, an important goal is that PFC be accepted by industry.

11.2 Limits of PFC Control

As with all techniques, PFC/PPC control has its limits and drawbacks as follows:

11.2.1 Multiple-Input, Multiple-Output Extension

The extension of PFC to 2MV/2CV is possible because the solution may be represented using a simple algebraic framework. There exists a place between multivariable, model-based predictive control (MBPC) and PID for a controller offering comparable, and sometimes superior, performance to multivariable control while preserving the "simplicity" of PID.

11.2.2 Constraints

The proposed solution should be a compromise that is acceptable for practical reasons and for ease of tuning. In some situations this approach may not be completely satisfactory.

11.2.3 Effort

It is necessary to make an effort to analyse and model a process, *e.g.*, evaporators. This is not always possible, either because the process is too complex to model quickly or the necessary expertise is not available for financial reasons, shortage of time and other resources. It should be noted that there will be a difference between the PID approach and that of model-based predictive control. This difficulty is a source for future research and development.

For these reasons, it is expected that predictive control will have only a limited number of application areas.

11.2.4 Risk

What are the risks associated with implementing this procedure?

As was previously discussed, most of the technical risk is connected with the quality of the model and it is imperative to verify the accuracy of the model's prediction *a priori*. In the rare cases where no modelling is possible, it is more prudent to resort to a technique that uses logical decisions only.

The real risk resides elsewhere, once the controller has been installed and commissioned:

- **The one-year test:** After one year of use, it is highly likely that all the functional requirements of the unit have been met, even the incidental cases. Some re-adjustments of the regulator may be required. Typically, when such problems do arise they are generally associated with the logical sequence of the continuous controller. The working conditions (process load) are taken into account in the model whose parameters (gain, dynamics, *etc.*) are rarely constant and vary with the operating conditions of the system. This property of passive, physical adjustment to the environment is fundamental to achieving robustness and is noted for its simplicity.

  If a first-principles model of the process is available this adaptation is easier. The number of industrial fields that have not yet adopted some form of model-based approach is falling. In fact, when a new first-principles model is accepted (commissioned) with its associated control strategy a whole body of knowledge may be explored.

- **The five-year test:** After five years "everything" has changed, raw materials, qualitative and quantitative production objectives, instrumentation, sensors and actuators. Furthermore, the person who installed the controller may have left the position without leaving adequate documentation. If the controller has not been maintained it is probable that its performance will have degraded significantly.

  How do you maintain such a degraded controller in these circumstances? In the case of PID, when it has proved necessary, the floor instrumentation personnel would adjust a few controls locally and find an acceptable performance. There would have been no need for any quality procedure with its associated documentation.

  In our case, a change in the local CLTR would suffice. But this approach does not represent the best professional practice.

Some major risks do exist:

- As predictive control is considered an "advanced" technique, people assume that the system is self-adaptive and requires no external supervision. The concepts of artificial intelligence, expert system and the self-tuning system may have been "oversold" by the vendor or just misunderstood. Whatever the reason, the "intelligence" of the controller should not be overestimated!

- Also, if an effective maintenance schedule has not been established, the operators will revert to using PID. A PID controller may be less effective but operators will be more familiar and comfortable with this technique and its maintenance procedures.

Implementing PFC as a standard block in a design package, with appropriate documentation supporting the model, is easy and may be incorporated using classical approaches (*e.g.*, IEC 61131-3). Consequently, it will be necessary to:

- Implement adaptive test protocols that will be systematically applied in order to gather data that permit process variations to be detected.
- Prepare a procedure for treating this data (filtering, analysis, *etc.*) that allows re-identification of the model with suitable identification methodology either onboard the PLC itself or on a remote computer.
- Present the results of the identification in such a way that personnel, other than the person who commissioned the controller, will be capable of updating the model and tuning the regulator.

The risks are not associated with the technique itself. Any potential problems that may arise will occur because of inadequate maintenance. In all cases, it is advisable to be prudent. Industrialists, who are not yet familiar with predictive control, will not accept one successful installation as a guarantee of success for production units located on their own sites. In fact, the impact of predictive control on industry as a whole has not yet realised its true potential – not for economical reasons, but for mostly cultural ones.

11.3 Final Remark

Past experience has left industry cautious of new techniques that appear to solve their problems, but ultimately do not meet their specific needs. As a consequence, industry is understandably slow to adopt new techniques. However, once the decision to adopt the PFC technique is taken, the time needed to learn, implement, operate, deploy and evaluate is short and its economic impact is practically immediate.

Appendix A

A.1 First-Order Process (K,T,D) in MATLAB®, C++ and VB

A.1.1 MATLAB®

```
%elem_PFC
%% Section 4.1, Figs 4.1 and 4.2
Target
%To evaluate the performance of PFC with respect to
%time delay compensation
%--------
tf=600;          % duration of test
w=1:1:tf;      % time
tsamp=1;       % sampling period ( seconds)
% process definition first order K T D
kp = 1;        % gain
Tp = 30        % time constant (sec)
np = 20;       % pure delay in sampling periods
%model definition
km = input('km (1) ');
Tm = input('Tm (30) ');
nm = input('nm (20) ');
%process
ap=exp(-tsamp/Tp); bp =1-ap; % difference equation
%model
am=exp(-tsamp/Tm); bm =1-am;
%CLTR  closed-loop time response (95%)
cltr = input('cltr (70) ');
h= 1;          %coincidence point here 1 for a first-order process!
lh=1-exp(-tsamp*h*3/cltr); %exponential reference trajectory
bh=1-am^h;     % ( free mode)
```

```
%constraints
MVmax = 120;              % max value of MV
MVmin =-10;               % min value of MV
DMV   = 5 ;               % speed limit of MV

%variables
MV   = zeros(1,tf);       % manipulated variable
CV   = MV;                % controlled varaible
sm   = MV;                % model output
pert = MV;                % disturbance
Sp   = 100;               %set-point
%------------------------------------
for ii=2+max(nm,np):1:tf
   if ii>300, pert(ii)=20;else,pert(ii)=0;end
   %process
   CV(ii)=CV(ii-1)*ap+bp*kp*(MV(ii-1-np)+pert(ii));
   %model with no dealy
   sm(ii)=sm(ii-1)*am+bm*km*MV(ii-1);
   % Prediction of process:
   % Initialisation of the exponential reference trajectory
   spred = CV(ii)+sm(ii)-sm(ii-nm);
   SFree = sm(ii)*am^h - sm(ii); %incremental free mode = sm(ii)*bh
   d= (Sp -spred)*lh+sm(ii)*bh;
   Sforced = bh*km;                 %base function is only one step
   MV(ii)=d/Sforced;
   % constraints
   if MV(ii)>MV(ii-1)+DMV; MV(ii)=MV(ii-1)+DMV;end % max speed
                                                   % constraint
   if MV(ii)<MV(ii-1)-DMV; MV(ii)=MV(ii-1)-DMV;end % min speed
       %constraint
   if MV(ii)>MVmax  ; MV(ii)=MVmax;end %max value of MV
   if MV(ii)<MVmin  ; MV(ii)=MVmin;end %min value of MV
end
plot(w,MV,'k',w,CV,'r',w,ones(1,tf)*Sp,'r',w,sm,'b')
grid
zoom
axis ([0 tf -25 155])
title( 'ELEMENTARY PFC /  model (b) / CV (r) / MV (k)')
text

%Note
%To be installed in PLC
%sm(ii)=sm(ii-1)*am+bm*km*MV(ii-1);
%spred = CV(ii)+sm(ii)-sm(ii-nm);
%MV(ii)=(Sp-spred)*k0+sm(ii)*k1; (ki being computed as parameters)
```

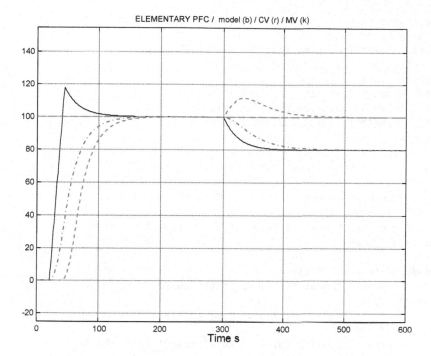

Figure A.1. Elemtary PFC control of a first-order process

A.1.2 C++

```
//Elementary PFC
#include <stdio.h>
#include <conio.h>
#include <math.h>

void main()
{
int i;
FILE *output;
        //After running, the file cp.dat is created in the current
directory, Visualisation of results with matlab
output=fopen("cp.dat","w");

//Variable declarations
//Process definition first order
float kp=1,Tp=30;
int np=20;
//Time, duration of test and sampling periods
float w[600],tf=600,tsamp=1;
//Model definition first order
float Tm=30,km=1;
int nm;//=20;
```

```
//coincidence point here 1 for a first-order process
float h=1;
//CLTR closed-loop time response (95%)
int cltr=70;
//constraints
float MVmax=120, MVmin=-10, DMV=5;
float lh,Sp,ap,bp,am,bm,bh,spred,sfree,sforced,d;
float pert[600],CV[600],sm[600],MV[600];
int max;
//Process definition first order
        printf("Process :Value of static gain?:");
scanf("%f",&kp);
        printf("Process : Value of Time constant?:");
scanf("%f",&Tp);
        printf("Process : Value of Delay time?");
scanf("%d",&np);

//Model definition first order
        printf("Model :Value of static gain?:");
scanf("%f",&km);
        printf("Model : Value of Time constant?:");
scanf("%f",&Tm);
        printf("Model : Value of Delay time?");
scanf("%d",&nm);
//process
ap=exp(-tsamp/Tp);
bp=1-ap;
//Model
am=exp(-tsamp/Tm);
bm=1-am;
//exponential reference trajectory
lh=1-exp((-tsamp*h*3)/cltr);
//free mode
bh=1-pow(am,h);
//Init
for ( i=0;i<599;i++)
    {
    CV[i]=0.0;
    MV[i]=0.0;
    sm[i]=0.0;
    pert[i]=0.0;
    }
//disturbance
for (i=0;i<300;i++) pert[i]=0;
for (i=300;i<600;i++) pert[i]=20;
Sp=100;
max=np;
if (nm>np) max=nm;
for (i=0;i<600;i++)    w[i]=i;
```

```
for (i=0;i<max;i++) fprintf(output,"%lf %lf %lf   %lf
%lf\n",w[i],MV[i],CV[i],Sp,sm[i]);


for(i=1+max;i<tf;i++)
    {
    //process
    CV[i]=CV[i-1]*ap+bp*kp*(MV[i-1-np]+pert[i]);
    //model with no delay
    sm[i]=sm[i-1]*am+bm*km*MV[i-1];
    // prediction of process: init of the exponential reference
    // trajectory
    spred=CV[i]+sm[i]-sm[i-nm];
    //incremental free mode = sm(i)*bh
    sfree=sm[i]*pow(am,h)-sm[i];
    d=(Sp-spred)*lh+sm[i]*bh;
    //base function is only one step
    sforced=bh*km;
    MV[i]=d/sforced;
    //constraints
    // Max speed constraint
    if (MV[i]>MV[i-1]+DMV) MV[i]=MV[i-1]+DMV;
    // Min speed constraint
    if (MV[i]<MV[i-1]-DMV) MV[i]=MV[i-1]-DMV;
    //max value of MV
    if (MV[i]>MVmax) MV[i]=MVmax;
    //min value of MV
    if (MV[i]<MVmin) MV[i]=MVmin;
                // In cp.dat    , first column = time, second column=
MV, third column = CV ...
    fprintf(output,"%lf %lf %lf   %lf
%lf\n",w[i],MV[i],CV[i],Sp,sm[i]);
    }
  fclose(output);
}
```

A.1.3 In Visual Basic

```
Private Sub ButtonStart_Click()
' stop simulation if you want to change process or model parameters
TextGainHP.Enabled = False
TextTetaHP.Enabled = False
TextRetardHP.Enabled = False
TextGainModelePFC.Enabled = False
TextConstanteTempsModelePFC.Enabled = False
TextRetardModelePFC.Enabled = False

'initialisation
SampleTime = 1                    ' sampling period (seconds)
'process definition first order K T D
Kp = CDbl(TextGainHP.Text)     ' gain
```

```
Tp = CDbl(TextTetaHP.Text)      ' time constant (sec)
Dp = CInt(TextRetardHP.Text)    ' pure delay in sampling periods
h = CDbl(TextHorizonPFC.Text)
' model definition
Km = CDbl(TextGainModelePFC.Text)
Tm = CDbl(TextConstanteTempsModelePFC.Text)
Dm = CInt(TextRetardModelePFC.Text)

MV = CDbl(TextSortie.Text)
Sp = CDbl(TextConsigne.Text)    ' set-point

MV1 = MV
SM = Km * MV
SM1 = SM
CV = Kp * MV
CV1 = CV
Sp1 = Sp
ReDim TabRetardmodele(Dm)
ReDim TabRetardProcess(Dp)
For I = 0 To Dm - 1
        TabRetardmodele(I) = SM
Next I

For I = 0 To Dp - 1
        TabRetardProcess(I) = CV
Next I

'constraints
MVmax = 120                 ' max value of MV
MVmin = 0                   ' min value of MV
DMV = 5                     ' speed limit of MV
OptionAuto.Value = True
Timer1.Enabled = True
End Sub

Private Sub Timer1_Timer()
time = time + 1
ImageJaugeC.Top = Sp
'CLTR   closed-loop time response (95%)
cltr = CDbl(TextCLTR_PFC.Text)
h = CDbl(TextHorizonPFC.Text)           'coincidence point: here 1 for
                                        'a first-order process!
'process
ap = Exp(-SampleTime / Tp)
bp = 1 - ap
'model
am = Exp(-SampleTime / Tm)
ah = Exp(-SampleTime * h / Tm)
bm = 1 - am
```

```
lh = 1 - Exp(-SampleTime * h * 3 / cltr) 'exponential reference
    'trajectory
bh = 1 - ah

    'Actual value for CV
    CV = TabRetardProcess(0)

    For I = 0 To Dp - 2
            TabRetardProcess(I) = TabRetardProcess(I + 1)
    Next I

    'Future value for CV (value for CV after the delay time)

    TabRetardProcess(Dp - 1) = TabRetardProcess(Dp - 2) * ap + bp *
Kp * MV1  '+pert(ii))

    EtiquetteMesure.Caption = Format(CV, "0.00")
    ImageJaugeM.Top = CV

    'SM :actual model output
    SM_Retard = TabRetardmodele(0)

    For I = 0 To Dm - 2
        TabRetardmodele(I) = TabRetardmodele(I + 1)
    Next I
    'model with no dealy
    TabRetardmodele(Dm - 1) = TabRetardmodele(Dm - 2) * am + bm * Km
* MV1
    SM = TabRetardmodele(Dm - 1)
    'drawing curves
    TracerCourbes
    'prediction of process : initialisation of the exponential
    'reference trajectory after the delay time
    SPRED = CV - SM_Retard + SM
 d = (Sp - SPRED) * lh + SM * bh
    If Km = 0 Then Km = 0.1
    Sforced = bh * Km              'base function is only one step
    If OptionAuto.Value = True Then
        'on est en auto
        MV = d / Sforced
        ' constraints
        If MV > MV1 + DMV Then MV = MV1 + DMV
                                ' max speed constraint
        If MV < MV1 - DMV Then MV = MV1 - DMV
                                ' min speed constraint
        If MV > MVmax Then MV = MVmax 'max value of MV
         If MV < MVmin Then MV = MVmin 'min value of MV
          TextSortie.Text = Format(MV, "0.00")
    End If
```

```
    'Save actual values
    CV1 = CV
    MV1 = MV
    SM1 = SM
    Sp1 = Sp

End Sub
```

A.1.4 First-order Integrator Process $H(s)=K/s(1+sT)$

```
%com_intg1er_4
% Section 4.6, Figure 4.4
clear all
close all
tf=1500;%test duration
w=1:1:tf;
tsamp=1;%sampling time
u=zeros(1,tf);e=u;sp1=u;sp=u;spp=u;...
sm1=u;sm2=u;sm=u;sm3=u;z=u;pert=u;v=ones(1,tf);
T=30;%process time constant
Kp=.05;%process gain
r=40;%delay process
am=exp(-tsamp/T);bm=1-am;
%MPC
theta=T*3; %decomposition process
as=exp(-tsamp/theta);bs=1-as;K2=Kp*theta*theta/(theta-
T);K1=(Kp*theta)-K2;K3=1;
cltr=input('cltr PFC (200) ');%time response MPC only tuning .!
h=(cltr/2);lh=1-exp(-tsamp*3*h/cltr);%coincidence
bmh=1-am^h;bsh=1-as^h;
MG=1;
for ii=2+r:1:tf
    %process
    if ii>650;pert(ii)=2;end
    sp1(ii)=am*sp1(ii-1)+bm*Kp*(e(ii-1)*MG+pert(ii));
    sp(ii)=sp(ii-1)+sp1(ii-r);
    %model
    sm1(ii)=am*sm1(ii-1)+bm*K1*e(ii-1);
    sm2(ii)=as*sm2(ii-1)+bs*K2*e(ii-1);
    sm3(ii)=as*sm3(ii-1)+bs*K3*spp(ii-1);
    sm(ii)=sm1(ii)+sm2(ii)+sm3(ii);%model
    spp(ii)=sp(ii)+sm(ii)-sm(ii-r);%prediction
    d=(100-spp(ii))*lh+sm1(ii)*bmh+sm2(ii)*bsh+sm3(ii)*bsh-
spp(ii)*bsh*K3;
    e(ii)=d/(K1*bmh+K2*bsh);
    if e(ii)>25;e(ii)=25;end
end
```

```
figure(1)
plot(w,e,'k',w,sp,'r',w,v*100,'k',w,pert*20,'b')
grid
zoom

axis([0 tf -10 110])
title('CONTROL OF INTEGRATOR + 1ST-ORDER')
text(550, 95, 'CV ')
text(850, 50 ,'Disturbance input x 20')
text(40, 30, 'MV')
```

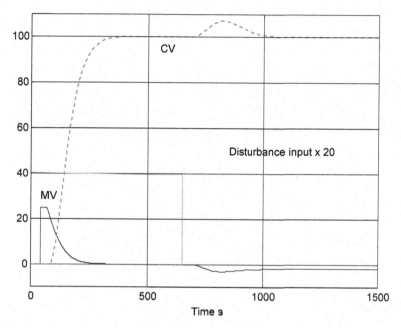

Figure A.2. PFC control of a first-order integrator process

Appendix B

B.1 Calculation of Heat-Transfer Coefficient for Water

```
% function watpar
% input for single pipe product or fluid of the heat exchanger
%        --> temp    : average Temperature [°C]
%        --> V       : Volume of pipe [m^3]
%        --> D       : diameter of pipe [m]
%        --> flow    : flow in pipe [m^3/sec]
%        --> lex     : length of pipe
% output
%        --> U       : heat exchange coef. for only one fluid [J / m^2Ks]
%        --> rhocp   : rho * cp for of the fluid [J / m3K]

function [U , rhocp] = watpar(temp , V , D , flow , lex);

%Re = Reynolds number
%Pr = Prandt number
%Nu = Nusselt number
k=0.023;
n=.8;
m=.333;


%%%% Polynom coeff
%-----------%heat exchange coef to be installed
aro = -9.6693e-008;
bro = 3.4622e-005;
cro = -7.0387e-003;
dro = 3.7265e-002;
ero = 1.0000e+003;

acp = 3.5169e-006;
bcp = -7.8602e-004;
ccp = 7.2444e-002;
```

```
dcp = -2.8623e+000;
ecp = 4.2185e+003;

amu = 4.2933e-011;
bmu = -1.0837e-008;
cmu = 1.0776e-006;
dmu = -5.6786e-005;
emu = 1.7836e-003;

alb = 7.6853e-010;
blb = -1.5530e-007;
clb = 5.7322e-007;
dlb = 1.9160e-003;
elb = 5.6103e-001;

%%% computation of parameters
rho = aro*temp^4 + bro*temp^3 + cro*temp^2 + dro*temp + ero;
cp = acp*temp^4 + bcp*temp^3 + ccp*temp^2 + dcp*temp + ecp;
mu = amu*temp^4 + bmu*temp^3 + cmu*temp^2 + dmu*temp + emu;
lambd = alb*temp^4 + blb*temp^3 + clb*temp^2 + dlb*temp + elb;
time = V / flow;
v = lex / time;
Re = rho*v*D/mu;
Pr = mu*cp/lambd;
Nu = k * (Re^n) * (Pr^m);
% output
U = (lambd*Nu / D);
rhocp = rho*cp;
```

B.1.1 Time-constant Calculation

```
X = qp/qf;
NUT = U*A /(rhop*cpp*qp); % Number of Transfer Units ( NTU)
z= exp( -NUT(1-X) );
if X==1;
   lambda= NUT/(1+NUT);
else
   lambda=(1-z)/(1-X*z);
end
   Cf=(rhof*cpf*Vf);  % fluid
   Cc=(rhop*cpp*Vp);  % product
   QT=(rhop*cpp*qp + rhop*cpf*qf);
   tau =( Cf + Cc + ( lambda / NUT / X ) * ( X*Cf + Cc   ) )
/QT;%time.cons
```

References

[1] Åström KJ, Hägglund T (2005) Advanced PID Control. ISA – The Instrumentation, Systems, and Automation Society, Research Triangle Park, NC 27709, USA

[2] Boucher P, Dumur D (1996) La Commande Prédictive. Éditions Technip, Paris

[3] Camacho EF (1993) Constrained Generalized Predictive Control. IEEE Trans.A.C, Vol. 38, N°2:327–332

[4] Clarke DW, Mohtadi C, Tuffs PS (1987) Generalized Predictive Control-Part I. The Basic Algorithm. Automatica, Vol. 23, N°2:137–148

[5] Clarke DW, Mohtadi C, Tuffs PS (1987) Generalized Predictive Control-Part II. Extensions and Interpretations. Automatica, Vol. 23, N°2:149–160

[6] Dittmar R, Pfeiffer BM (2004) Modellbasierte prädiktive Regelung. Oldenburg

[7] Culter CR, Ramaker BL (1980) Dynamic Matrix Control. A Computer Control Algorithm. Proceedings JACC, Paper WP5–B

[8] Garcia CE Prett DM, Morari M, Model Predictive Control: theory and practice – a survey. Automatica, (1989) Vol. 25, N°3:335–348

[9] Grimble MJ (2001) Industrial Control Systems Design. Wiley, New York

[10] Landau ID (2002) Commande des Systèmes. Éditions Hermès, Paris

[11] Maciejowski JM (2002) Predictive Control with Constraints. Pearson Education Limited, England

[12] Qin J, Badgwell T (2003) A Survey of Model Predictive Control Technology. Control Engineering Practice S:733–764

[13] Rawlings J, Meadows E, Muske K (1994) Nonlinear model predictive control: a tutorial and survey. In Proceedings of IFAC ADCHEM, Japan

[14] Richalet J (1993) Pratique de la Commande Prédictive. Éditions Hermès, Paris

[15] Richalet J (1993) Industrial Application of Model Based Predictive Control. Automatica, Vol. 29, N°5:1251–1274

[16] Richalet J, Eguchi H (2007) Model Predictive Control – Principles and Application of PFC. Nihon Kogyo Shuppan, Japan (In Japanese)

[17] Richalet J, Rault A, Testud JL, Papon J (1978) Model Predictive Heuristic Control: Applications to industrial processes. Automatica, Vol. 14, N°5:413–428

[18] Rossiter JA (2003) Model Based Predictive Control – A Practical Approach. CRC Press Boca Raton

Index

Other titles published in this series (continued):